Children of the Universe

Hoimar von Ditfurth

Children of the Universe

THE TALE OF OUR EXISTENCE

Translated from the German by

JAN VAN HEURCK

Atheneum *New York*

1974

The publisher wishes to express thanks for the use of the following photographs: No. 1, World Wide Photos; No. 15, Metta Holst; Nos. 20b and 20c, E. B. Stüdli; No. 23, Air Albrecht Brugger, Stuttgart (released by the government of Baden-Württemberg).

The line drawings are by Heike Meibaum.

Originally published in German as *Kinder des Weltalls: Der Roman unserer Existenz.* © 1970 Hoffman und Campe Verlag, Hamburg.

Library of Congress catalog card number 73-91629
ISBN 0-689-10588-6
Published simultaneously in Canada by McClelland and Stewart Ltd.

Manufactured in the United States of America by
Quinn & Boden Company, Inc., Rahway, New Jersey

Designed by Kathleen Carey

First Edition

Translator's Note

The English text owes a great deal to the efforts of Peter Stetson, my "science consultant," who spends much of his time in the astronomical observatory at Wesleyan University, armed with telescope and computer, trying to coax stubborn stars into betraying their whereabouts. Fortunately for this translation, Mr. Stetson came down from among the stars long enough to help me with the English rendering of scientific vocabulary and concepts. We made a few minor additions which we thought might prove helpful to English-speaking readers.

I must also thank fellow-translator Krishna Winston, who read over the completed manuscript looking for sunspots and other flaws, and came away, like the rest of us, properly star-struck.

I hope that this translation conveys something of the imagination, awe, and occasional whimsy of the original. Strictly speaking, this is not merely a book of popular science. Underlying the scientific data is a philosophical premise and even an evangelical intent. Von Ditfurth suggests that our discovery of the *physical* unity of the cosmos can help cure our culture of metaphysical *Angst* and restore the sense of wholeness we lost a little after Eden. This seems to me rather a tall order. After all, the mere ubiquity of physical law does not seem much of a substitute for the "Love which moves the sun and other stars." It must always be difficult for human beings to relate to a mere machine, however vast and smoothly functioning. But strangely enough, von

Ditfurth achieves on the mythical or imaginative plane what he may fail to achieve on the didactic: He actually "animates" or ensouls the vast machinery of the universe.

What seems to me to set this book apart from many science books of our day is precisely its, in the literal sense, "animated" character. While translating it, I often found myself recalling the medieval bestiaries and early works of natural history from the Greeks to the Renaissance. For in *Children of the Universe*, as in these earlier works, science has not yet been completely divorced from allegory, from theology, from fable, and from magic. (It was impossible to include in the English edition the often whimsical running heads. In one of them, for example, the author describes the earth's collisions with meteors as battles with "giants," as if our world were still the arena of Norse mythology.)

Nature here is often personified, and always has implications for human life. Not only the different branches of science but the spectrum of human emotions and institutions are mirrored in these pages, which provide us with a sort of *speculum mundi*, a compendium or mirror of the world in the medieval sense. Just beneath the surface of twentieth-century scientific fact we find the preoccupation with immortality; the ancient concepts of "microcosm" and "macrocosm"; and echoes of the "As above, so below" of hermetic magic. The "model stars" and mental games of the astronomers resemble acts of mythic imagination. Sober as the facts may be, underlying them is von Ditfurth's recognition that science is "merely the continuation of metaphysics in another form." By the end of the book, as we participate in the dawn of life or the private life of a star, we have returned for a time to an older, a pristine world, when "there were giants in the earth" and "when the morning stars sang together." It is almost the landscape of Genesis.

JAN VAN HEURCK

Contents

Children of the Universe

The Scene of Action

We cannot really conceive what is meant by a phrase like "three billion years." The history of life on earth encompasses a stretch of time so vast that it defies our understanding. By comparison, the *spatial* dimensions of the stage on which this history has unfolded seem tiny. The oldest known Pre-Cambrian fossils (primitive one-celled organisms, similar to blue algae, that are found in a remarkably ancient shalelike stone of northern Michigan known as the "gunflint chert"), and perhaps recent South African finds as well, prove that life has existed on earth for at least three billion years. Such a stretch of time staggers the imagination. Our experience of time and our ability to judge and visualize distances (including distances in time) are determined by our biological constitution—a constitution which permits us to move no more than six to ten miles an hour under our own steam, and affords us a life span of a mere seventy or eighty years.

So we cannot conceive what is meant by the phrase "three billion years." This may be one reason why many of our educated contemporaries still cannot accept the Darwinian concept of evolution. They find it inconceivable that a seemingly blind process of selection, operating among random mutations, could "in the brief span of a few billion years" cause one species to develop out of another. Nor can they imagine how the struggles for existence within a given species should have produced the multiplicity of highly sophisticated life forms we observe today.

The evolutionary process is indeed "inconceivable," although in a far more literal sense than its detractors may intend. Because of our physical makeup, we inevitably underestimate time periods of great magnitude. The anti-Darwinians fail to take into account our built-in misperception of time. Unfortunately, even our method of writing numbers supports our unconscious tendency to underestimate large measurements. Our ingenious decimal system makes it easy for us to deal with large sums. In going from 1,000 to 1,000,000, we simply add three more zeros. As schoolchildren, we are never taught that the addition of three more zeros involves a refined sort of "numerical shorthand"; actually, in order to make 1,000,000, the number 1,000 ought to be written out no less than a thousand times.

There are a few familiar tricks, crutches for our imagination, which can give us a fleeting notion of the quantities we are dealing with. By saying one number every second, a person can count to 1,000 in about a quarter of an hour. Counting at the same pace for an eight-hour day, he would need a whole month to reach 1,000,000. It would take a whole lifetime, allowing eight hours a day and one second per number, to get to one billion. The results are just as startling if these time spans are translated into spatial terms. If we represent the course of natural history during the last three billion years as a model three meters long (about ten feet), then the whole history of mankind, including the prehistoric epochs since the Bronze Age, would not take up enough space to be visible except under a microscope. On this scale, 10,000 years would occupy only 1/100 of a millimeter (= ten microns).

These and any other mental exercises we might perform simply prove one point: We cannot picture the aeons which have elapsed in the evolution of life on earth. It is miraculous enough that we have been able to measure them. We have invented numbers and names for vast quantities of time and space; but our minds cannot grasp their reality.

Seen against this backdrop of time, the space we occupy—our "theater of action"—looks quite tiny. This theater is the surface of our earth. To be sure, the "stage" measures approximately 197,000,000 square miles in all, including the oceans and the polar regions. (When examining the total history of life on earth, we must consider the oceans, the jungle, and other regions which are hostile to human life but do support other forms.) Of course, an area of 197,000,000 square miles is nothing to sneeze at. But it is still something our imagination can grasp without too much difficulty. We can picture a square with sides just over 13,760 miles in length. The area of this square would correspond roughly to the earth's surface.

Photographs brought back from the recent flights of the astronauts have revealed to us how the earth looks from outer space. We can see clearly how minute the stage is on which the drama of human history has unfolded; we can take it in at a glance. This lost-looking sphere hanging there in the infinity of space has served as the "scene of action" throughout those "inconceivable" three billion years of life. Everything human beings have ever felt and thought, suffered or accomplished, and all that went before—from the union of the first spontaneously generated organic molecules and their infinitely slow development toward the first biological system capable of reproduction, to the "invention" of photosynthesis, the rise of the multicellular organism, the movement from sea to land, the development of warm-blooded life, the birth of consciousness and, most recently, of the capacity for self-reflection—all this has unfolded on the surface of this sphere which measures a mere 7,920 miles from top to bottom.

To get a complete picture of our stage, we must take a third dimension into consideration. The stage we are discussing is not a level area 197,000,000 miles square, but an area which *rises* from this surface. How does this third dimension, height, affect our scene of action? We might set the altitude of the stage at 6,500 feet. Of course, man can survive at even twice this altitude without an oxygen mask. But in order to maintain normal body functions above the 6,500-foot limit, several weeks of adjustment would be necessary. Changes would have to take place in the composition of the blood: chiefly an increase in the number of oxygen-bearing red corpuscles. Besides, it is questionable whether regions lying above 6,500 feet may still be considered part of the ecosphere, the area inhabitable by life. The number of life forms occurring above this limit is so infinitesimal that we need not consider them here. In any case, the average height of our planet's land masses comes to only 2,700 feet. Therefore, 6,500 feet is a generous limit to set. It is irrelevant for our purposes to discuss the few areas above this level.

So the stage we are trying to portray has an altitude of about 6,500 feet. But what about its depth? After all, we do not plan to confine ourselves to human life: our stage is the scene of earthly life as a whole. We must also look at the seas—the more so since all present indications are that water was the source and cradle of life.

Taking the oceans into account will not, however, greatly affect our calculations. Even if we make ample allowance for them, they will add only 3,250 feet to the depth of our stage. Of course, living creatures exist in the sunless depths below this level. In recent decades oceanographers have established that fairly abundant animal life inhabits even the deepest regions of the ocean floor—at depths of more than 32,000 feet. And these deep regions (unlike the high mountainous regions) represent a far larger proportion of the earth's surface than is generally supposed. About two-thirds of the earth's total surface (about 139,000,000 out of 197,000,000 square miles, or 71 percent) is covered by water. But 82.7 percent of the area covered by water lies at that level which oceanographers designate as "abyssal depths" (between 6,500 and 19,500 feet).

In the context of this discussion, however, we are justified in limiting the depth of the ecosphere to 3,250 feet. Our justification lies in the special qualities of deep-sea fauna. Given the variety and abundance of life forms in general, deep-sea fauna must be regarded as a highly specialized, relatively uniform type of animal life. Systematic probes with deep-sea nets have shown that even such life as there is thins out rapidly at depths of 3,200 to 6,500 feet. But there is another reason for considering a lower limit of 3,250 feet more than ample: All deep-sea creatures discovered up to now originated in the upper water levels, at a depth of only 1,600 to 2,000 feet. This is the limit to which sunlight can at least partially penetrate under favorable conditions.

The possibility of photosynthesis (and therefore of basic food production) ceases well above this depth. Therefore, any life found below these levels must have originated in the upper levels irradiated by the sun; only later did it migrate to the deeper levels, after having passed through a gradual and difficult process of adaptation. Moreover, this deep-sea life is dependent for survival on the organisms that exist in the upper, sunlit levels of the sea and at death filter downward in a steady, nourishing stream.

Therefore we may say that the scene of action has a vertical spread of only 9,750 feet. But in order to get a clear idea of what this means, we have yet to translate this quantity into the scale of Illustration 1, the astronauts' photograph of the earth suspended in space. In this photograph the globe has a diameter of about 12 centimeters (4.7 inches). On this scale, the ecosphere would have the thickness of exactly .03 millimeters (1/1,000 inch), forming a film so thin as to be invisible to the naked eye.

So this is all we have. This is the space in which we live, the space where all of human history has unfolded; it is also the only space provided for life in any form we know. One hundred ninety-seven million square miles across, 9,750 feet in height (or depth), and the whole thing, as it were, wrapped around a sphere 7,920 miles thick. The interior of the sphere, beyond the first few hundred miles, must be classed among those regions considered inimical to life. The sphere itself is suspended in an immeasurably vast and inconceivably empty space. This, then, is where we live, where the drama of human history plays itself out.

Now that we have seen what all these numerical relationships imply, what can we conclude about our situation on this earth?

If we look at our earth without preconceptions, it seems to offer a perfect paradigm of an existence as precarious as it is improbable. Viewed objectively, the reality of our earthly existence directly contradicts the sense of permanence and security instilled in us by our immediate environment. Let us back away a little from our everyday surroundings—to about the distance from which the astronauts took that photograph of our planet shown in Illustration 1. We shall then see that our habitual perception of the world as stable and reliable is an illusion. The facts not only fail to support this illusion; they stand in grotesque contrast to it. A tiny area of light, warmth, and life, wrapped paper thin over a ball hanging forlornly in empty space: This is how our world suddenly looks to us.

From a distance of just a few thousand miles, everything we daily take for granted is transformed into a mere pinpoint in space. This pinpoint all at once looks so alien to us; the existence of our earth and the life on it appears utterly inexplicable. Thus we readily accept the popular notion that our planet represents a unique event in the cosmos, the product of a whole chain of bizarre accidents. And in fact this conclusion may seem unavoidable. When we first attempt to view our surroundings objectively, the new perspective can be overwhelming. No wonder that the image of our home as a forlorn planet lost in space has long been a permanent component of the so-called "modern" view of the world.

Of course, such an image has not always been prevalent. Up until a few centuries ago, man regarded himself and the earth as part of a cosmic order. To be sure, even medieval man made a distinction between the earthly realm lying "beneath the moon" and the heavenly spheres of the fixed stars which hover in eternal immutability over this sublunar world of mortal beings and transitory things. But he did not for a moment doubt the basic unity of these two cosmic realms; nor did he doubt that the border between them was spanned by a dense network of powers and influences constantly interacting with each other. These powers revealed themselves to men through an abundance of images, allegories, and mythological concepts.

The world view we now designate as medieval did not survive the advent of the scientific method. With time, myths and metaphors proved too imprecise for the human mind. Moreover, these myths and metaphors were already losing their viability because human beings were taking literally what at first had been meant only as an image.

With the discovery of the scientific method, men lost the illusion of a familiar, secure world which had its place in an all-encompassing cosmos. The disillusionment proved far more radical than anyone could have foreseen. What brought Galileo before the tribunal or Giordano Bruno to the stake was not simply the stubborn intolerance of a Church immobilized by its own dogma. It is too easy to make this sort of accusation; and above all it is unjust. We too readily overlook the very real fear that underlay the Church's aggressive reaction to scientific analysis. Today we are so accustomed to the scientific approach that we fail to understand the extraordinary shock it must have produced in the contemporaries of Galileo and Bruno.

The first scientists began to paint a picture of the world somewhat like the one we have been painting here. When man attempted to free himself from his habitual attitudes in order to see the world "as it really is," he was confronted with a terrifying possibility: the possibility that the cosmos might in fact be both indifferent and unrelated to him. Thus the seed was sown for that "modern" view, still current today, that the earth with everything on it is dangling in the isolation of a universe whose cold majesty disdains it.

Today we have long since become used to the thought of our humble position in the cosmos. Deep down, we are probably even proud of the detachment with which we accept our "true" situation. We have resigned ourselves to being marooned in a universe which is infinitely large and infinitely dead. Of course, pride in our objectivity is probably not the only emotion we experience when we think about our place in the scheme of things. Clearly, natural science (despite all notions to the contrary) does not consist solely in the gathering and classification of facts. The facts are a means to an end. In the final analysis, natural science is simply part of man's attempt at self-understanding. Therefore, the conviction that we are isolated in a vast, empty, and hostile cosmos—a belief founded in scientific fact—has had far-reaching emotional consequences. I should like to suggest, although such a contention could never be proven, that much of the cynicism and nihilism characteristic of the modern psyche can be traced to this chilling conception.

Scientists are now discovering this world view to be essentially false. What takes place in space a few thousand miles above our heads is anything but meaningless for human lives. What happens elsewhere in the universe relates far more closely to us and our environment than any of the earlier mythologies had supposed. Perhaps the most intriguing, certainly the most significant, insight of modern geology and astronomy is that everything is interlaced: the microcosm with the macrocosm, and events in front of our noses with events at the limits of the perceptible universe.

The new findings which are revolutionizing our understanding of ourselves form the proper subject of this book. These findings are coming to us from the most diverse disciplines of natural science, and not necessarily from the most modern. Geophysics and paleontology are contributing as much as space research and cosmology. Amazingly enough, it has proved possible to merge systematically the insights of these disparate disciplines to form a single image of the cosmos and of our place in it. The very unanimity of these insights suggests that we are witnessing what may prove to be a major turning point in our understanding of the world.

As we examine this new and still incomplete image of our world, we shall encounter some baffling and unexpected relationships. For example, our lives depend on a frail force barely strong enough to pull a compass needle toward the north. It is probable that this force, the earth's magnetic field, would not exist if the earth had no natural satellite. Apparently our earth would be uninhabitable without the moon. What more telling symbol could there be of how closely our sublunar world is bound up with the powers that lie outside it? We shall also see that a constant exchange of matter is going on throughout the universe; our earth is included in this process. We now know that virtually every one of us has at one time held in his hand a stone which came from the moon, if not from regions even farther out in space. Most recently, we have even found indications that this cosmic exchange of matter decisively influenced the course of evolution. So the possibility exists that we would not be what we are today but for this phenomenon, only now coming to light. This cosmic exchange alone would be enough to refute our old view that the universe has no bearing on human life.

But this is only one of many examples. Most people will be equally surprised to learn that the sun, our earth, the moon, and all the other planets of our solar system represent a sort of "second generation" of heavenly bodies. There will be other intriguing discoveries: (1) With the sole exception of pure hydrogen, all known matter (including the matter from which we ourselves are made) must have originated in the center of stars, through atomic nuclear fusion reactions; (2) cosmologists have found out that it can get dark at night only because the universe is *not* infinitely large; (3) even the spiral shape of our Milky Way has turned out to be a basic precondition for our own existence.

To many people all this may at first sound not merely unfamiliar, but unbelievable; they may think I am exaggerating for the sake of effect. But it is all quite literally true. These discoveries represent the work of thousands of scientists over a period of decades. These scientists amassed so much data that at first it seemed impossible to perceive an underlying order. But now the picture is changing. Certain key discoveries are functioning like basic crystals around which many other hitherto unrelated facts are beginning to group themselves. A fascinatingly new yet strangely familiar image is suddenly emerging. It is the image of a universe whose parts are all closely related. The laws and forces ruling this universe justify our calling it a "cosmos": a harmonious entity whose every part is responsive to every other.

This is a far cry from the picture of our earth traveling forlornly through a nightmare of empty space. Our sublunar world of mortal beings and transitory things is actually linked in a thousand ways with the starry spheres. The depths of space are ruled by the same laws and forces which govern terrestrial life, and we ourselves would not exist if our earth were really as isolated from those depths as we have long tried to persuade ourselves. A planet capable of sustaining life did not come into being independently of the rest of the universe. On the contrary, the scene of action is molded by influences which reach us from the farthest corners of the universe.

The universe, then, is not that empty, cold, and lifeless space whose majestic indifference once inspired us with fear as well as admiration. It is the living soil from which our earth has sprung;

and our planet is joined to it by a thousand roots. We are not some tiny oasis permitted to exist in a hostile universe solely because of our insignificance. Instead, our earth can be shown to be a focal point where various cosmic powers conjoin to fashion a living world.

The scene of action thus resembles a stage in two respects. It is a stage in that the history of life is played out upon it. But what exists on the earth's surface is supported, much like a troupe of actors, by countless backstage assistants. These "stagehands" form a network that extends far beyond the theater walls; and they generally remain hidden from the spectator unless he knows where to look for them. Certainly the earth is not the center of the universe. This illusion has been discarded forever. But this crowded earth *is* a focal point in the universe: one of those perhaps innumerable places in the cosmos where both life and consciousness could flourish. Many factors united to produce and maintain the right conditions. What a concentration of mighty forces upon one more or less tiny point! The details of this process and of its discovery form the topic of this book.

Science and Man's Self-Image

NOWADAYS scientific publications printed more than twenty or thirty years ago are considered out of date. Because of the rapid advances being made in fields like astronomy and physics, most publications become obsolete even sooner. There is always a possibility that new research will supersede scientific texts during the brief interval between their writing and their publication. For this reason many scientific periodicals have adopted the practice of printing the date of composition in parentheses after the title of an article. And since a book may easily have become obsolete by the time it is printed, many scholars are unwilling to undertake the several years of work involved in compiling a new textbook. Thus new textbooks and manuals are long overdue in a number of modern scientific fields. The old textbooks are simply reprinted from time to time, brought up to date by a summary of recent research. This state of affairs prevails not only in certain areas of medicine and biology; remarkably enough, it prevails in the field of astronomy as well.

Except for professional astronomers and astrophysicists, virtually no one is aware that astronomy too has been caught up in the ever-increasing tempo of scientific research. Astronomy, which along with medicine is the oldest of the sciences, developed over the centuries at a very leisurely pace. It grew out of the patient and tenacious observations and calculations of men who were obsessed with the mystery posed by the night sky. This mystery, seemingly accessible to both the telescope and the naked eye, nevertheless surpassed all understanding. The early astronomers made slow and laborious progress. Despite all their efforts they usually had to content themselves with passing on their data, in the hope that the following generations of astronomers could make sense of what now seemed obscure. But these early astronomers could afford to disregard their frustrations, for in a few cases they actually succeeded in finding answers to their questions. They knew that it was possible to discover meaningful relationships between remote cosmic phenomena and the laws operating here on earth.

The progress of astronomy has now accelerated more rapidly than any outsider can imagine. More discoveries have been made in this field during the past ten years than in all the centuries since Copernicus announced that the earth was *not* the center of the universe. In recent years, discoveries in astronomy have been practically treading on each other's heels. Significantly, one publisher recently included several footnotes in the fourth edition of a well-known astronomical manual (published 1967); these footnotes pointed out those few details in the book which still corresponded to the newest research findings!

In view of the snail's pace that characterized the history of astronomy, such a phenomenon seems almost grotesque. Up to now, astronautical technology has accounted for only a small portion of the new findings. No doubt the tempo of discoveries in astronomy will increase even more as soon as the first unmanned observatories begin to orbit the earth outside its atmosphere. Above all, our knowledge will increase when a fixed observatory can be established on the moon, whose almost total lack of atmosphere will afford us an uninterrupted view into space.

Many people still fail to appreciate the tremendous efforts the present generation of scientists is expending in order to achieve goals such as this moon observatory. Some of our contemporaries shake their heads whenever they think of the billions of dollars being spent in the attempt to realize our scientific potential in space. Whenever the subject comes up, they begin to calculate how many schools could be built for the cost of a single moon shot, or how many miles of highway could be developed with what is spent on one false start at Cape Kennedy or Baikonur. However realistic and convincing these protests may sound, the people who make them fail to recognize what is at stake here for humanity and its future development. They are provincial in the true sense of the word; their view reflects a narrow horizon. This must be said even of some well-known personalities who are wholly competent in their own fields. These people have not awakened to the crucial significance of discoveries in astronomy. This significance does not lie in their practical results, even though these results will no doubt far outstrip our expectations. Rather it lies in a fact which even supposedly educated people almost always overlook. That fact can be stated thus: The prevailing image of the world at any given time always forms the basis for how human beings view themselves.

This underlying connection between man's view of the world and his view of himself is so fundamental that we often overlook it. Every age has its ideological framework—the meaning people attribute to their existence, which influences the way they conduct their lives. But the underlying meanings are obscured by thick layers of political and historical realities, specific events, accidental constellations, and deep-seated traditions. These layers resist all our efforts to penetrate them. It is not an easy thing to uncover the thought patterns unique to each age. In any case, we are not concerned here with the validity of any particular interpretation of human existence or with the possible criteria for judging a given interpretation. Nor need we consider whether basic attitudes may be said to evolve or progress from one historical epoch to another. In the context of this discussion, only one thing need concern us: There is a relationship between man's evaluation of the world and his evaluation of himself. The behavior of whole generations is automatically determined beforehand by the way they perceive their world and their own role in it.

In coming decades men will have to come to grips with the fact that all the available surface of the earth is crammed with human beings. How will they relate to each other? How will they feel about themselves and their lives when they start bumping into each other at every turn? How they react will be determined by the discoveries present-day scientists and astronomers are making about our world. More specifically, the behavior of our children and our children's children will depend on their image of the world and of their place in it; and this image will be derived from scientific findings which are being made at this very moment.

Anyone who has been sensitized to the relationship between world-image and self-image will require no further proof that space probes, moon stations, and new earth observatories (especially larger radio telescopes) are more important than anything else we can construct today. They are more important than schools and new highways, necessary as these things undoubtedly are. We often hear the objection that we should "solve our problems right here on earth" before squandering "astronomical" sums just to transport a few tons of machinery to the inhospitable surface of other heavenly bodies. This sensible-sounding remark lacks foresight. Scientists are now making decisions that will decide the future course of history. What can we do but shake our heads when we hear people talk of the irrelevance of space research? Indeed, it is amazing that some years ago a Nobel prize winner could have made the often-quoted remark that space travel is merely the expression of a meaningless drive to achieve new technological records—a triumph of intellect but a deplorable failure of judgment!

Actually just the opposite is true. How can we "solve our problems right here on earth" before we have inquired what our true role in the universe should be? Our situation is growing ever more crucial and chaotic. How can we hope to establish order in the world before we know what the world really is? What standard should we use to judge by? Of course there are some people who cannot feel the fascination of intellectual adventure; they experience no joy in penetrating new, undreamed-of aspects of the reality that surrounds them. But even these people must realize the importance of a photograph like that of the earth shown in Illustration 1.

This photograph transcends all political, linguistic, and ideological boundaries; it shows every citizen of earth something about our true situation, something that nothing else could show him in such a concrete way. Heretofore we have seen the earth only as the stage of rivalries and conflicts. This photograph affords us a fresh, more accurate view of all that we have in common. In the long run, such a new perspective could provide far more help and more hope than the endless debates of politicians, statesmen, and committees.

Of course this photograph is a rather crude example for the point I am trying to make—the argument that money spent on space research represents a senseless extravagance is more superficial and ill-considered than it sounds. The desire to set technological records is insufficient to explain the astronomical sums and the unprecedented human effort which have gone into putting a single astronaut into orbit for a few days. We would have to go back to the building of the pyramids and the Great Wall to find a historical parallel for such a mammoth undertaking.

The photographs of the surface of Mars relayed back to earth by the American interplanetary space probe Mariner IV are without doubt the most expensive pictures ever made. But what underlies this expense, and all the military, political, and nationalistic arguments used to coax the needed funds from recalcitrant legislatures? What underlies them is a phenomenon which ought to arouse both awe and hope in us. The fact is that this generation, so often scorned for its materialistic and nihilistic attitudes, has the strength and élan to mobilize all its potential for the sake of an intangible goal. Our generation is ready and able to commit all its resources to extending the horizon and the consciousness of humanity.

To be sure, we live in an age of technology. But technological man has many visions. Some people still bemoan the fact that science now plays the leading role in our cultural life. Such people have not yet recognized that science simply represents the continuation of metaphysics in another form.

The remarkable and even revolutionary discoveries in astronomy during the last ten years must be seen against the background of their meaning for us, our self-image, and our attitudes toward life. So far no clear pattern has emerged from these new discoveries. Although they involve objectively measurable data, their implications still appear enigmatic, even inexplicable. The strongest impression we glean from these data may be simply that our old world view is tottering. That old view is yielding to an unfamiliar image of the universe—a universe which, far from dwelling in timeless immobility, is filled with powerful forces engaged in endless development. We are included in the history of this cosmos; its laws determine our earthly surroundings and the very core of our being. We shall now examine some of the discoveries which reveal our indissoluble ties with the rest of the universe.

Astronautics and the Dimensions of Space

For centuries we have known that this solid ground under our feet is really a sphere spinning through empty space, just one among countless heavenly bodies. At least we have paid lip service to this fact. But we have not yet fully accepted its emotional implications. In time we will feel the full impact of the Copernican world view. Future historians may one day regard this experience as the quintessential event of our culture. Invisibly but inexorably, our new attitudes will shape the thought of all succeeding generations. Life on earth in no way represents a unique or privileged phenomenon. Scientists have always accepted this fact; in time it will be universally acknowledged. We will abandon the naive and anthropocentric illusion that the earth alone has given birth to life.

This illusion has had a disastrous influence on human history.

It has bred some dangerous assumptions. Many people regard life on earth as the pinnacle and goal of all creation. Despite the real dangers which threaten it, they assume that its preservation is somehow magically guaranteed. Such attitudes are largely unconscious—and thus all the more dangerous. Most people agree that it would be an awful thing if human violence were to vent itself in a major nuclear explosion. On the other hand, they secretly feel that civilization could never be permanently destroyed.

Human aggression is instinctual, like hunger. No instinct can be dammed up forever. Sooner or later a starving man will begin to rummage through garbage cans. During polar expeditions, men have boiled shoe soles to extract their protein content. When a basic drive is frustrated, it eventually seeks expression through another channel. A man whose aggressive drive has been thwarted reacts very much like a man who is starving: The tension builds up inside him. That is, his "threshold of aggression" becomes very low. Now the slightest incident can unleash a disaster. But when the threshold of aggression is comparatively high, only a powerful stimulus can produce aggressive behavior. Where such a high threshold exists, diplomats often succeed in smoothing over serious international incidents like border violations or local incursions.

Historians are still racking their brains to find the "true" causes of World War I. The two political assassinations in Sarajevo must have deeply shocked a society that still functioned according to feudal mores. But how could the act of a few fanatics have triggered off a war that engulfed the whole civilized world? For half a century scholars have been puzzling over this question; they have delved into innumerable archives. Yet they have found no rational explanation for a holocaust that resulted in the death of millions. But those of us who have studied general biology, congenital behavior patterns, and hereditary trigger mechanisms think we understand the riddle. The war was a manifestation of instinctual human aggression.

It is high time that we acknowledge the instinctual nature of aggression. We must learn to see beneath the surface causes of present-day world tensions. We all read the political press and listen to the daily news broadcasts; we hear about conflicts of interest between various nations. But these conflicts do not adequately explain the universal state of crisis in the world. Man has acquired the power of "overkill"—the ability to destroy all life on this planet. We have the potential to achieve a self-induced apocalypse. But the effects of absolute annihilation seem to outweigh its causes. Our tangible conflicts of interest hardly seem to warrant the destruction of the world. Actually, we are not involved in a simple confrontation between conflicting interests; we have been caught up in a reciprocal escalation of anxiety and aggression. At bottom, this anxiety and this aggression are manifestations of instinctual behavior. The diplomats dealing with international problems have been trained in foreign languages, international law, and the rules of protocol. It might have been more useful if they had devoted their time to social psychology and the study of cultural behavior patterns. But perhaps there may still be time for them to learn.

No instinct can be permanently repressed. Modern man tends to regard aggressive behavior as a sort of anachronism, an embarrassing remnant of his less civilized past. But no instinct can be simply willed out of existence. Therefore our fate depends on our ability to recognize the relationship between instinct and behavior. To control behavior, we must first understand its roots. Space research can help us to view our surroundings more objectively—with reason rather than instinct. The study of space has shown us that our earth is really too small for civil war. Space research indeed seems more important than all other "problems we should solve right here on earth," for it can aid us to solve these other problems.

Space research has revealed another important fact about our earth. The potential for life is not unique to our planet. And no law of nature guarantees that our form of life will survive. We ourselves must determine whether or not we survive. We are in danger of self-extinction. If we ignore this fact, we are simply increasing the danger.

Our generation has learned that the earth is probably only one of countless inhabited planets in the cosmos. We are a mere local variation; our fate will not affect the evolution of life forms elsewhere in the universe. This discovery should help free us of dangerous illusions. Our world does contain some seeds of hope. One day all men may recognize that they are members of a single community—citizens of the earth.

Since the launching of our space probes, we have all seen how small the earth and its role in the cosmos really are. Men have ventured a short distance beyond the earth's atmosphere. But how insignificant our space flights seem, compared with the immensity of space itself!

It makes one dizzy to think of all that space engineers have accomplished since the launching of the first artificial satellite, Sputnik I, on October 4, 1957. They have tackled and solved baffling technical problems, although nothing in the past history of science could give them a clue to the solution of these problems. Yet seen against the backdrop of space itself, their magnificent achievement seems puny and absurd. Our space engineers are justifiably proud of putting an astronaut into orbit around the earth or landing him on the moon. But a true science of astronautics or cosmonautics is still a thing of the future. Up to now we have merely set up a sort of "local transit line" in the neighborhood of a little planet called earth. No man has yet succeeded in leaving this little neighborhood.

Let us try to construct a mental scale model to give us an idea of the dimensions of space. First, we might shrink our solar system to 1/100,000,000 of its present size. Now the earth looks like a grapefruit about 4½ inches in diameter. On this scale, the earth's surface seems almost completely smooth, rather like a polished billiard ball. To be sure, on our model Mt. Everest has a height of 3/1,000 inch; if we touched the mountain with our fingertips, we might feel a slightly roughened spot on the polished ball. If we breathed on this ball, the thickness of the film made by our breath would exceed the depth of the oceans! Now let us add some features to our model. We must picture the moon as a ball about 1½ inches across. This ball is circling the earth at a distance of some 12½ feet, describing an orbit with a total diameter of 25 feet. The whole earth-moon system could be placed in a room measuring 25 by 25 feet.

But now comes the first giant step in our calculations: In this scale model, the sun is a mile away! With a diameter of about 46 feet—about 1½ times the height of a high-diving board—the sun would not fit into an ordinary building. Mars, our neighboring planet in the solar system, is about 2¾ inches across. But even when Mars comes closest to the earth (i.e., when the earth overtakes Mars "along the inside track"), the red planet is still some 1,640 feet away. To reach the outer edge of our solar system, the orbit of Pluto, we would have to make a fatiguing two-day march across our scale model. The 2-inch sphere of Pluto is about 37 miles away! Now let us place our astronauts inside the same scale model. All our manned flights have orbited the grapefruit-earth at a distance of only about 1/10 inch above the surface. In the foreseeable future our astronauts will journey no farther than they have between 1969 and 1973, the 12½ feet from the earth to the moon.

The term "space flight" is still just wishful thinking; and it is sheer nonsense for us to speak of having "conquered" space. We have not yet even set foot in outer space itself. Recent discoveries indicate that true "outer" space begins beyond the orbit of Mars or even the orbit of Pluto—that is, outside the limits of our solar system. Within our solar system, conditions are largely determined by electromagnetic and corpuscular radiation emitted by our sun. Therefore the area around our planets cannot be considered typical of interstellar space.

The radiation in our solar system has profoundly influenced

living conditions on earth. In a later chapter we will examine the relationship between the earth and cosmic radiation. For now, let us return to the concept of "space travel"—a phrase which people throw around far too lightly. We must try to get a more accurate idea of the true dimensions of space.

Now back to our mental scale model (scale 1:100,000,000). We have a 46-foot-wide sun orbited by a grapefruit-sized earth a mile away. Now how far away is the nearest star—the nearest "neighboring" sun? Our nearest neighbor, Alpha Centauri, is so far away that we would need to become astronauts inside our own scale model just to reach it. This fiery ball the size of our own sun is actually as far away as the "real" moon is from the earth! The term "outer space" applies only to the interstellar distances separating the various suns or stars. (The terms "sun" and "star" mean the same thing; the former designation of stars as "fixed stars" is now obsolete, since stars do in fact move.) We now have to picture some 100 billion balls of fire like Alpha Centauri, averaging 46 feet in diameter, more or less equidistant from one another, and held together by mutual attraction in a single so-called "galaxy." (Of course, we are back to our old problem: We can write numbers like 100,000,000,000, but we cannot picture the reality they symbolize.) A galaxy is an independent stellar system, lens- or discus-shaped, which rotates around a single center of gravity. (See Illustrations 2 and 3.)

Many people are more familiar with another, inaccurate term for galaxy—"spiral nebula." This misleading term, like much scientific vocabulary, reflects the past history of science. "Galaxy" is the correct technical term for certain types of large, relatively independent conglomerations of stars. But not all galaxies are spiral in nature. Moreover, the use of the term "nebula" with reference to external galaxies (i.e., galaxies outside our own) is archaic. For centuries scientists observing galaxies through their telescopes believed that they were actually looking at gaseous clouds, called "nebulae." (Many true nebulae do exist in our galaxy.) But in the 1920s the American astronomer Edwin

Powell Hubble, using his new giant telescope (over eight feet in diameter), observed our neighboring "spiral nebula" in the constellation Andromeda. Hubble observed that the supposed "cloud" was actually composed of individual stars. True nebulae are composed of gas. Thus spiral galaxies, which are made up primarily of individual stars, are not true nebulae.

All the galaxies ever photographed (whether spiral or of other shapes) have a blurred or "nebulous" appearance. This cloudy look is misleading: it results from our imperfect photographic techniques, which distort the true appearance of the galaxies. The average galaxy is composed of from 20 to 100 billion suns or stars. But even when we look through our most powerful telescopes, these separate stars look like tiny dots. Thus it is extremely difficult to photograph a galaxy. The photographer allows the light from the galaxy to accumulate on a photographic plate. The stars emit a very weak glow, so that their light must accumulate for several hours. But the light-specks striking the film from the stars of a galaxy are actually smaller than the individual grains or crystals composing the film. In order to register the presence of light on its surface, an entire grain of film must "change" or "turn": that is, the grain will record the star's light as if this light occupied an entire grain of film, although in reality it occupies only a fraction. It is impossible to photograph details finer than one grain. Because of this "limit of resolution" —the inability to make visible the separate parts of a galaxy—we continue to see these largest of all cosmic structures as indistinct clouds.

Our photographic techniques give us a false impression of the distances obtaining within galaxies. But another scale model may help correct this impression. Let us imagine that it is beginning to rain, or rather to drizzle, all along the vast U.S. highway system. Each raindrop is at most 7/1,000 of an inch in diameter. On an average, the raindrops fall some fifty miles apart. One drop falls in New York City; the next lands fifty miles north, in Connecticut; the third in Massachusetts; the fourth in Vermont, and so on. These drops are the individual suns or stars. The intervals between the drops correspond to the intervals between the stars in a galaxy. From 20 to 100 billion such "star drops" make up a single galaxy, or "milky way."

The expression "milky way," like "spiral nebula," is historical in origin. External galaxies are called milky ways only in a very figurative sense, when a scientist wishes to stress the similarity of the external galaxies to our own. Apparently all the stars in the universe belong to one of the innumerable galaxies outside our own Milky Way Galaxy. Some of these galaxies are visible through our telescopes. Viewed from the earth, the various galaxies occupy different positions in space. A galaxy may appear to us as a rounded spiral, as an ellipse, or (from a side view) as a flat lens. (See Illustrations 4, 5, and 6.)

Astronomers have calculated the orbits of various stars; they have learned that some stars are describing highly eccentric orbits. One day these orbits will carry the stars out of our galaxy. Thus the space between individual galaxies may not be entirely empty of stars. It may well be populated by suns which have lost contact with their home systems and now pursue an isolated course. (And how strange it would be if some of these suns had given birth to planetary systems and even inhabited planets! What would become of these planets and their life forms when their sun severed its normal ties with a galaxy?) However, these eccentric suns are exceptional cases which need not concern us here. Generally speaking, stars are not equally distributed throughout the universe; but neither are they scattered at random. They are all members of one of the many galaxies. Space approaches being "empty" only in between the galaxies themselves.

Now let us see where our own sun fits into this picture. Our sun is a star like all others; like many stars, it belongs to a spiral galaxy. (We shall soon see that in fact not all galaxies have spiral arms.) Our sun differs from other suns in only one respect: It is ours. This statement is not an invitation to join in

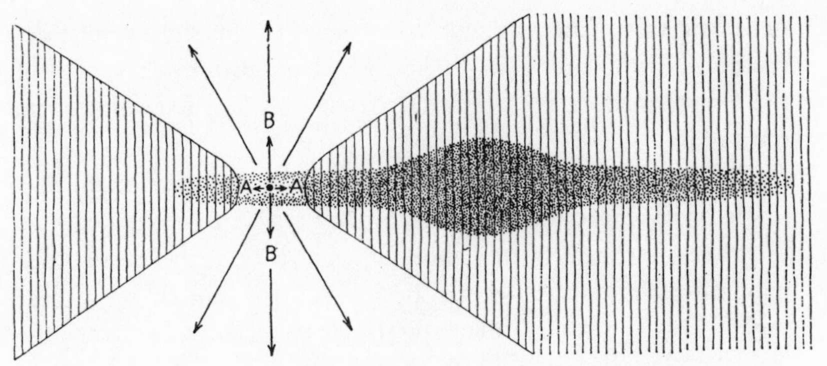

Side view of our own galaxy (shaded area). The dot marks the location of our solar system. Alphabetical designations are explained in text. The striped areas represent areas of space invisible from the earth; they are clouded over by cosmic dust. We have an unrestricted view into outer space only above and below. Our field of vision has an X or hourglass shape.

some sort of cosmic local patriotism: It simply points out that we view our own galaxy from a particular point in space. We can view other galaxies from the outside; we see our own from the inside. Therefore we do not see our galaxy as an extended ellipse or a lens-shaped formation. Our sun is just one of the 100 billion stars composing our galaxy; from this vantage point we cannot see the spiral pattern. Many people do not understand this fact when they look into the night sky. Seen from within (that is, from just *one* of the planets belonging to *one* of the suns that compose it), a spiral galaxy appears as a grouping of stars in the form of a band or stripe—the "Milky Way." The diagram above shows how our galaxy appears to us on earth. (Also see Illustration 7.)

The central dot marked in this diagram represents our solar system. Let us assume that I am standing at this point and looking out at the night sky. I turn my eyes along a line parallel to the plane of our own spiral galaxy (along line A). In this direction I see the stars composing the Milky Way Galaxy; they are thickly massed beside and behind each other. If I shift my gaze above or below this plane, the number of stars in the sky rapidly decreases. Next I look along a line *perpendicular* to the plane of the lens-shaped galaxy (line B). Along this line I see almost no stars at all.

The more distant members of our galaxy are distributed in a single band all around the earth, a band visible to the naked eye. However, our nearer neighbors appear to be more evenly distributed over the sky, as anyone can see for himself on a clear night. This is how our own spiral galaxy looks to us "from the inside." (See Illustration 7. In this photograph the brightest stars actually form spiral arms. This proves that, viewed from another galaxy, our Milky Way Galaxy would have the typical spiral form which characterizes all other spiral galaxies. In recent years astronomers have applied special radioastronomical techniques to the study of our galaxy; they have established supplementary proof that it does in fact have the typical spiral shape.)

Even in ancient times, men were struck by the brilliance of the white band of stars in the sky: They endowed this band with mythical properties. It became known as the "Milky Way." In the middle of the eighteenth century Immanuel Kant began to think about this systematic grouping of stars around the earth. His brilliant analysis of our Milky Way Galaxy is still well worth reading. Kant theorized that the Milky Way was actually a galaxy viewed from the inside, and that the various large elliptical "objects" visible in the sky were independent milky ways, or "island universes" similar to our own. At this time the traditional term "Milky Way" remained in common use. Little more than forty years ago, Hubble discovered the true nature of the spiral and elliptical "cloud" formations known as "nebulae": They were actually independent stellar systems lying far outside our Milky Way Galaxy. In structure and composition, the independent stellar systems were basically identical to our own galaxy.

From then on, the term "milky way" (our word "galaxy" comes from the Greek *gala-*, *galaktos*, for "milk") was sometimes loosely applied to all those formations formerly designated as spiral "nebulae." This phrase nicely topped off the verbal confusion. Now both terms—"galaxy" and "milky way"—mean more or less the same thing. Both designate a type of large independent stellar system. But "galaxy" is the scientifically correct term.

Galaxies are the largest formations in the universe. Their dimensions stagger the imagination. Even our scale model (the widely spaced raindrops) could not really help us to picture the distance separating two stars. So how could we possibly picture the dimensions of an entire galaxy? It would take around 100,000 years for light to travel from a point on one side of our galaxy to a point directly opposite it on the other side. Therefore our galaxy has a diameter of 100,000 light-years.

The cosmic swirls of stars composing a galaxy are inconceivably vast, so vast that even the oldest of them have rotated only about twenty times around their own axes in all the 10 to 15 billion years since the universe began! And yet these galaxies have been rotating at such speed that their swiftest-moving suns cover more than 300 miles per second. (In our own galaxy, the region of fastest rotation is near halfway between the center and the edge. Our own solar system travels at only about 162 miles per second. The previous diagram shows our position in the galaxy.) The following mental experiment offers us a clue to the dimensions involved:

Let us imagine that we have stuck a pin into the picture of the galaxy shown in Illustration 5. Now let us enlarge the galaxy to its true size. The pinhole has made an enormous hole in the galaxy. Even if we traveled in a spaceship moving at the speed of light (186,000 miles per second), we would not live long enough to get from one side of the hole to the other! Of course no real spaceship could ever travel at the speed of light. But even a mythical vehicle like this would take at least 700 years to reach the other side of the hole. We can learn something else from this useful game, invented by Eduard Verhülsdonk. This pinhole in the galaxy would destroy around one million suns! Astronomers estimate that some 50,000 of these suns would have their own planetary systems. Of these 50,000 solar systems, at least several hundred would be inhabited by some form of life. All this represents a mere pinpoint within a single galaxy. The galaxy in turn is no more than a pinpoint in the universe. How presumptuous it is of man to speak of having "conquered" space!

The story of Nova Persei can also help us to understand the dimensions of space. Nova Persei revealed some startling facts to astronomers. On February 21, 1901, a new star, a so-called "nova," was discovered in the constellation of Perseus. Nowadays a nova is no longer considered an unusual phenomenon. Astronomers have observed several hundred similar cases. A nova (or the rarer supernova) is not literally a new star: it is an already-existing star which has undergone a dramatic transformation. A nova results from the explosion of a star which up to this point has been quite "normal." Suddenly the star undergoes an enormous atomic explosion during which some of the star's matter is ejected into space. A nova loses at least 1/10,000 of its mass; a supernova may lose up to ten percent.

Illustration 8 shows the so-called Crab Nebula, a gigantic expanding cloud. This famous formation lies in the constellation of Taurus. It constitutes the remains of a supernova explosion observed by Chinese astronomers in A.D. 1054. Now, almost a thousand years later, the cloud is still expanding. Modern astronomers have photographed it each year and compared the results. Each year its diameter increases by a tiny amount: only .21 second of arc. The expanding cloud lies 4,000 light-years from the earth. Astronomers have determined that its particles are still hurtling away from the center at speeds exceeding 620 miles per second! This seems an incredible rate of speed. After all, over 900 years have passed since the explosion was first observed.

But then astronomers discovered Nova Persei, which made even the Crab Nebula look tame. After the original explosion of the nova, the cloud continued to expand. Astronomers measured its expansion from week to week. Illustration 9 shows the original nova as it looked in 1901, the year of the eruption. The astronomers eagerly computed the rate of expansion of the exploding cloud. They proceeded in the usual way, finding out how far the cloud lay from the earth and measuring its continual increase in diameter. For once they had succeeded in locating a nova which was still in its preliminary stages. Therefore they expected to find that the cloud was still expanding at unusually high speeds. But the rate of expansion was not merely unusual; it was impossible! The astronomers repeated their calculations. They used spectroscopic and photometric techniques to measure the distance between the cloud and the earth, and repeatedly computed the cloud's increase in size. But their findings never varied: The exploding cloud of Nova Persei was expanding in all directions at the speed of light itself!

The astronomers were sure that their calculations could not be correct, since at that time they already knew that matter can never attain the speed of light. (Even if this speed had been theoretically possible, a nova would not have time to build up such speed; a nova explosion lasts for only a few hours or days.) The hydrogen atoms or the protons and electrons in the expanding cloud were still matter in a gaseous form; therefore they could not travel at the speed of light.

Some time passed before the riddle was solved. At first astronomers continued to assume that they had made some error in measurement. We now know that the measurements performed in 1901 were entirely correct. The solution was simply this: What the astronomers saw radiating from the center of the explosion was not matter at all, but light! This explanation has since helped astronomers to understand similar cases. The nova happened to explode inside a gigantic interstellar cloud of very fine cosmic dust. What the astronomers had actually observed was the light of the explosion traveling through the dust at 186,000 miles per second. As the light moved through the cosmic dust, it turned the dust into a brilliant cloud.

The story of this nova gives us an idea of the dimensions of interstellar space. Light travels faster than anything else in the universe; yet if we had been watching Nova Persei in 1901, its light would have seemed to creep across the sky. In the same way, a racing car on a distant horizon seems to move like a snail. The creeping light was inside our own galaxy, almost in our own backyard. Nova Persei is "only" about 3,000 light-years from the earth. Even so, its light would have seemed to move at a snail's pace.

Let us take another look at this neighbor of ours. If we could see our Milky Way Galaxy from the outside, it would look just like the Andromeda Galaxy (Illustration 5). On this photograph, our earth and Nova Persei would be only 5 to 6 millimeters apart (0.19 to 0.23 inch). But the corresponding distance in space is so vast that even light must travel 3,000 years to cross it.

Even light requires millennia to cross the space between us and our nearest galactic neighbors. Thus whenever we look at the starry sky, we are always quite literally looking at the *past*. The stars lie at varying distances from us. The light from one of these stars may reach our eyes in fifty or sixty years. Another star may appear to lie right beside the first, yet its light has been traveling thousands of years to reach us. We see each of these stars just as it was when the light set out on its journey. A star may have been destroyed centuries ago in the explosion of a supernova. Then an astronomer comes along and begins to investigate this nonexistent star; he may photograph it and subject it to spectroscopic analysis. The star may be no more than a thousand light-years away; yet the news of its death will not reach us for a thousand years! The Crab Nebula (Illustration 8) represents the remains of a supernova explosion observed by Chinese astronomers in A.D. 1054. But the star did not really explode in 1054; the actual explosion took place around 3,000 B.C. That is, the supernova was located about 4,000 light-years from the earth, so that it took 4,000 years to transmit the image of the exploding star to our planet.

Most people are aware of this "time lag" when they observe the stars. But they may not understand the full implications of the patterns they see in the night sky. The stars above us seem to lie side by side in space; some of them appear to be linked to form famous constellations. But these stars are not "simultaneous" in time. Even within a single constellation, we are looking at stars which existed in many different epochs of the past; often they existed millennia apart. We frequently overlook an important fact: We can never see our own galaxy exactly as it is (or was) at any individual moment in time! The farther we peer into our Milky Way Galaxy, the farther we are seeing into the past.

We still see nearby stars as they looked years or centuries ago. We see the far side of our galaxy as it was about 80,000 years ago. Thus the various regions of the galaxy appear to us in a distorted form; they are distorted by time. The degree of distortion depends on their distance from the earth. This fact does not affect the validity of scientific data regarding our galaxy; during the brief span of 80,000 to 100,000 years, the basic structure of a galaxy changes very little. Nevertheless, this visual "time lag" gives us something to think about; it provides a concrete example of the indissoluble relationship between space and time.

Our image of our own galaxy is invariably distorted by time. Paradoxical as it may sound, this rule does not always apply to other, far more distant galaxies. For example, let us look at the galaxy in Illustration 4. This galaxy is situated in such a way that we must look down on it from above. In a case like this, all the stars composing the galaxy are more or less equidistant from us; from our vantage point they are seen to lie in roughly the same plane. Since they are approximately equidistant from us, these stars we perceive as side by side in space are also simultaneous in time. (That is, stars which lie at varying distances from the earth occupy different time zones; but stars more or less equidistant from the earth occupy roughly the same time zone.) When we observe other stellar systems, the time lag is never as pronounced as when we are looking at our own. Even if these stellar systems are inclined at unusual angles in relation to our earth, this rule still applies. Our view of other galaxies can never be as distorted as our view of our own. When we view another galaxy from the side, the stars are densely massed; they prevent us from seeing very far into the galaxy. Thus most of the stars we see are more or less the same distance away, and therefore simultaneous in time. Moreover, when we observe a galaxy from the side, along its equator, our view is barred by masses of cosmic dust. This dust, which appears in every galaxy, also prevents us from seeing very deeply into the stellar system. The dust itself is very fine. It appears dark because it forms particularly dense concentrations at the very edge of the galaxy, that is, along its equatorial plane. Illustration 6 demonstrates this phenomenon quite clearly.

Our view of our own galaxy is especially distorted because we see it from the inside. That is, we can see stars at various distances from the earth in all directions. These stars are not equidistant from us; therefore they are not simultaneous in time. Of course, we cannot see our entire galaxy. If we could, then the stars we see would vary even more in their distance from us; that is, our galaxy would appear even more distorted. As it is, our field of vision is surprisingly narrow. Astronomers sometimes perform systematic star counts in order to compare the density of stars at the center of the galaxy with the density at its edge. At distances of up to 6,000 light-years from the earth, these star counts can give us some idea of the density of a few types of stars, but only in a few directions. For certain types of stars, the counts are not accurate within thirty light-years, and they become inaccurate for more types of stars as distances increase. One of many limiting factors involved in the observation of stars within our own galaxy is cosmic dust. Although this dust is very thin, it becomes perceptible over long distances. When we look along our galaxy's plane of rotation, the dust appears particularly dense.

We confront a very different situation if we look out along a line perpendicular to the plane of the Milky Way. In this direction, our stellar system is thinning out; therefore the cosmic dust is thinnest here too. Straight up and straight down—in these two directions we have an almost unobstructed view of "extragalactic" space. The diagram shown earlier in the chapter clearly illustrates this fact. Until recently we knew nothing about the small, X-shaped area of our Milky Way Galaxy directly perpendicular to the earth. Nor did we understand the structure of our galaxy as a whole. During the past twenty years, the development of radio astronomy has opened many doors to astronomers. The interstellar hydrogen in our galaxy emits radio waves; these waves are particularly concentrated in the center of the galaxy, simply because in this area we are dealing with more hydrogen. Unlike light rays, radio waves easily penetrate cosmic dust; they enable us to observe nearly our *entire* Milky Way.

This same cosmic dust was once responsible for an extremely interesting and instructive error on the part of astronomers. For almost two hundred years, their rejection of a theory of Immanuel Kant prevented astronomers from understanding the true nature of spiral and other sorts of galaxies. In March of 1775, Kant dedicated a book to his sovereign, Frederick the Great. This book was the *Universal Natural History and Theory of the Heavens*, in which Kant logically defended his theory about the roughly oval, dim, white "patches" which the men of his time had so often observed through their telescopes. Kant believed the patches to be independent stellar systems lying far beyond our Milky Way. We can still take pleasure in reading how the brilliant citizen of Königsberg simply and logically proved a theory that seemed outrageous in his day. We now know that every step of his argument was correct. Yet his theory was not accepted until almost two hundred years later. As already mentioned, in 1923 Hubble demonstrated that the bright border regions of the

Andromeda Galaxy actually consisted of separate stars; the supposed "nebula" was not a cloud at all.

The diagram shown earlier can help us to understand one reason why astronomers refused to accept Kant's theory. Opponents of the "extragalactic" theory of spiral "nebulae" had one very impressive argument in their favor. From the earth, the "nebulae" appeared to be systematically arranged around one area of our Milky Way. Around the turn of the century, hundreds of these formations were photographed and their positions entered on star charts. The charts indicated that the spiral systems were not equally distributed throughout space. Their number increased as one approached the "galactic pole," the line designated B on our diagram. By the same token, the formations thinned out toward the plane of our Milky Way until they disappeared altogether (direction A).

Astronomers could draw only one conclusion from the unequal distribution of the mysterious elliptical objects in the sky: The arrangement of the "nebulae" was somehow related to our Milky Way. Then they reached a second conclusion which seemed to follow logically from the first: The formations must be "intragalactic" objects, outlying bodies of our own stellar system. Otherwise the formations would not have carefully arranged themselves around our galactic pole. Looking back now, we can easily detect the flaw in this seemingly airtight argument. The first step in the argument was quite correct. The distribution of external galaxies—their concentration in the area visible along our galactic pole—*is* in fact related to our Milky Way Galaxy. More precisely, it is related to one aspect of our galaxy, the cosmic dust. The uneven concentration of cosmic dust in our galaxy obscures our view of external galaxies. We have a clear view of space only along the galactic pole (line B in the diagram); as a result, we see external galaxies only along this pole. In reality, galaxies are fairly evenly distributed throughout the universe.

Since the development of radio astronomy, even cosmic dust

no longer bars our view of the heavens. But we can still see very little with the naked eye. Even on a clear moonless night, no more than 6,000 stars are visible to the naked eye. Without a telescope, even the Milky Way looks like a cloudy, palely glowing band. Of course, all the stars we can see belong to our galaxy; in fact, they all lie very near our own solar system. In other words, with the naked eye we can see less than $1/10,000,000$ of the 100 billion suns contained in our galaxy!

There is one exception to this rule. We *can* see one "extragalactic" object with the naked eye. On a moonless night, a person with good eyesight who knows just where to look can barely make out a small dim speck. This speck lies in the constellation of Andromeda, far beyond the Milky Way Galaxy. The speck lies at an inconceivable distance from the earth. We have tried in vain to picture vast distances with the aid of scale models and mental games. But the heavenly bodies in our scale models lay within our own solar system or at least within our own galaxy. Even intragalactic distances are so vast that we cannot see our own galaxy as it really is; its various regions are always distorted by time.

We are about to take a giant leap into space: The pale speck in the constellation of Andromeda lies outside our own galaxy. It is the Andromeda Galaxy, 2,000,000 light-years away. We used the enlarged photograph of this galaxy (Illustration 5) when we performed Verhülsdonk's "pin-sticking" experiment. We have just encountered our first example of *inter*galactic distance. The speck of light in Andromeda represents an independent spiral galaxy, a milky-way system like our own.

Now we are really getting acquainted with "outer space." The Andromeda Galaxy is our closest galactic neighbor. The more deeply our giant modern telescopes penetrate into space, the more galaxies they discover. Illustration 10 shows a whole group of galaxies. Each approaches the size of our own Milky Way system, being composed of from 50 to 200 billion suns. They lie so far from the earth that they appear slight and delicate.

Illustrations 11 and 12 show larger groups of galaxies, with very few foreground stars.*

But we have not yet arrived at the limits of the visible universe. Special photographic techniques have supplied us with pictures showing several thousand galaxies, or milky ways—on a single photograph! This means that right now our instruments can detect several billion milky ways. Unfortunately, our investigations are limited by the range of our instruments. But we have no reason to suppose that we have already discovered all the galaxies. Astronomers believe that the total number of galaxies in the universe far exceeds the number of suns composing each of them (from 50 to 200 billion).

But the number of galaxies, vast as it may be, is probably not infinite. Recent discoveries in astronomy suggest that we may soon know something definite about the limits of the cosmos. To be sure, we must take care not to confuse the limits of our present knowledge with the limits of what there is to be known. Scientists have repeatedly made this mistake, and are probably making it still under various disguises. Many findings which now appear valid may conceal some pervasive error. We ought to remain skeptical. All the same, it does appear that the universe may be much smaller than we once believed. In fact, a large portion of it may already lie within range of our instruments.

We have already noted some effects of the intimate link between space and time. As distances increase, the effects of this link become ever more apparent. Up to now we have discussed the space-time link only in relation to *intra*galactic distances. The diameter of our galaxy amounts to a mere 100,000 light-years. We have seen that our nearest neighboring galaxy, the Andromeda Galaxy, lies twenty times farther away than this. Most of the galaxies shown in Illustration 12 are several hundred million light

* "Foreground stars" are in our own galaxy. These stars lie between us and any distant object we may try to photograph. When astrophotographers take pictures of objects outside our galaxy, some of the foreground stars always creep into the picture. They are usually overexposed.

years away; the light from the most distant of them has been traveling a billion years to reach us.

When we look at this photograph (Illustration 12), we are already looking at the very ancient past. We are now dealing with billions of years, that is, we are dealing with spans of time which are meaningful in terms of cosmic evolution. The most remote heavenly bodies we have been able to photograph lie some 3 billion light-years away. But the antennae of modern radio telescopes receive impulses from objects 6 to 8 billion light-years away.

Anyone who read the last sentence attentively may have noticed that I did not refer to these distant cosmic "objects" as galaxies or milky ways. This was quite intentional. Our investigations are based on light or radio waves. But the data they bring us stem from the very remote past. The objects in question must sometimes be almost as old as the universe itself. In recent years, scientists have received some bizarre and myserious data relating to these objects. These mysterious data may have something to do with the age of the universe. For years astronomers have been recording findings which seem to contradict each other and the laws of physics. A physicist would describe these findings as "incompatible with the known laws of nature."

These mysterious phenomena lie at the limits of the perceptible universe. Light or radio waves traveling from these "objects" take at least 6 to 8 billion years to reach us. Astronomers will be occupied with these riddles for many years to come. But they have already roughly determined the age of the universe. The data being received from the most remote objects in space approach the age of the universe itself. This fact may explain their strange and contradictory qualities. There is one hypothetical conclusion we might draw about why these objects now appear so mysterious. Are we really so far off the track if we imagine that *other laws* governed the universe when it was young?

We have another reason for being concerned with the problem of where the universe ends. Let us ask ourselves an apparently simple question. The profound implications of this question did not become apparent until the beginning of the nineteenth century. At that time Wilhelm Olbers, a physician and astronomer of Bremen, asked this question too. It sounds straightforward enough: Why does it get dark at night? Simple as the question may sound, astronomers labored for more than a hundred years to find a satisfactory answer.

Olbers reasoned this way: If the universe were infinitely large and if there were as many stars in the rest of the universe as there are in the part visible to us, then there would have to be one star at every point in the heavens. Of course their light would seem dimmer the farther away from us they were. Nevertheless, there would have to be an infinite number of stars piled up behind each other at every point—*if* the universe were infinite. In this case the entire night sky would have to be as bright—and as hot!—as our sun. For this reason Olbers posed the question: Why does it get dark at night?

Modern astronomy found the answer to this question. The answer sounds as simple as the question; but it is just as meaningful. The universe is *not* infinitely large. Every day we experience the fact that night falls when the sun sinks below the horizon. This commonplace experience is one of the proofs that the universe, vast though it may be, cannot be *infinite*. Once again we see how closely our everyday world is related to the laws of the cosmos.*

We began this chapter by examining our solar system; now we have traveled across interstellar space. One heavenly body concerns us more than all others: the third among nine planets moving around a star. But we must view our earth against the background of the universe. Our sun is only one among 100

* Some modern cosmologists have devised various ways to reconcile the famous "Olbers' paradox" with the often-cited theory of the "expansion of the universe." In the views of these scientists, both theories basically arrive at the same conclusion.

billion stars composing a stellar system known as a galaxy, or "the Milky Way." This galaxy in turn rotates in a space so vast that it can house countless billions of such galaxies.

Everything I have just outlined is established fact. It is a reality which surrounds us all our lives. In order to see this reality we need only go outside on a clear moonless night, away from bright streets and houses, and look up at the sky.

Our Spaceship Earth

WE HAVE been trying thus far to put together a picture of the universe surrounding our earth. The earth may seem like a tiny particle of dust traveling forlornly through the vastness of space. In relation to its surroundings, our planet is indeed minuscule. But in other ways it can hardly qualify as "forlorn." An invisible but indestructible network of gravitational forces holds the earth on its course. This network also includes the other heavenly bodies of our solar system. We have all heard it said that gravitation causes heavenly bodies to describe curving orbits, and that the centrifugal force generated by these orbits balances the pull of gravitation, i.e., of centripetal force. This description suggests that two separate and opposing forces are involved in holding the earth on a stable course. In reality, gravitation alone maintains the planetary orbits. Literally speaking, "centrifugal force" does not exist; it is properly regarded as a fictitious or "pseudo-force." That which we call centrifugal force is merely a useful illusion that helps us treat certain complex situations as simple ones. (In this case, the *rotating* frame of reference of our solar system is treated as an *inertial* frame.) In any case, gravitation has held the earth on a stable course for billions of years.

As it travels along its course, the earth resembles a spaceship. Its "crew" consists of the entire animal kingdom, including man. This may seem nothing more than a fashionable poetic metaphor, but in reality the metaphor is more literal than we suppose. Like a spaceship, the earth carries with it everything it needs to support the life on its surface. For example, part of its cargo consists of the oxygen required by human and other animal life.

The parallel between the earth and a spaceship may become clearer if we consider this fact: The supply of oxygen on the earth is limited just as it would be on a spaceship. More than 3 billion human beings, not to mention all the other animals, are now consuming oxygen. The earth's atmosphere contains an enormous supply of oxygen, but even this supply will not satisfy the needs of all earth's creatures for more than 300 years. In one respect, the earth has a distinct advantage over a spaceship. If a meteor happens to strike a spaceship, it can create a hole. The ship may begin to lose oxygen which can never be replaced. Something like this actually happened on the perilous flight of Apollo 13. Like a spaceship, the earth cannot replace oxygen once it has been lost. But, fortunately, the earth cannot spring a leak. The gravitational field prevents our oxygen from leaking into space. Despite this advantage, the earth's oxygen supply will not indefinitely satisfy our needs. Animals and men will exhaust it in a mere 300 years.

The day is coming when our oxygen supply will run out and the entire crew of Spaceship Earth will be threatened with suffocation. Various technological and industrial processes are cutting short the little time we have left. We are about to be presented with the bill for our centuries of oxygen consumption. The situation looks grave—for a moment. But then we are bound to stop and think: Can this be a true description of our earth? After all, for billions of years there has been enough oxygen in our atmosphere to satisfy the needs of all the living creatures on the planet. To be sure, our population has increased; and modern industrial processes have stepped up the rate of oxygen consumption. But these factors do not adequately explain why our oxygen supply should suddenly be running low.

There is an answer to the riddle. Animals and men *will* use up the earth's oxygen supply in 300 years. But meanwhile the plants are breathing too. They continually renew the oxygen in the atmosphere, enabling us to use the same oxygen over and over again.

Plants do not merely *renew* our oxygen supply; they actually created it in the first place! The earth's primitive atmosphere contained virtually no oxygen. It was rich in methane, carbon dioxide, ammonia, hydrogen, cyanogen, and various other gases we now regard as poisonous. Strange as it may seem, this was the ideal atmosphere for the development of life. The basic building blocks of life were complex molecules of high molecular weight. Free oxygen would have immediately oxidized these molecules, destroying them as fast as they could form. In the beginning oxygen was a deadly poison which might have prevented life from developing on this planet.

Certain molecular combinations developed the ability to reproduce themselves. These were the first "living" organisms. One billion years after life began, the earth's surface was covered with a solid carpet of plants. These plants caused a worldwide revolution, affecting all existent life forms. With the aid of sunlight, plants could build complex organic molecules like carbohydrates, fats, and proteins out of simple inorganic compounds. This special ability gave plants an advantage over earlier life forms. Plants can utilize the sun's energy in performing their metabolic functions. The capacity for photosynthesis still distinguishes the plant from the animal kingdom. Animals cannot achieve photosynthesis; thus they depend on plants to supply them with the needed organic compounds. To survive, all animals must feed on plants; or they must eat other animals which feed on plants.

Biochemists do not completely understand the complex process of photosynthesis. But they do know one thing: The metabolic processes of plants produce an important waste product—oxygen. The capacity for photosynthesis gave plants a natural superiority over other forms of life; they swiftly overran the earth. At the same time, they gradually increased the oxygen content of the atmosphere. At that time, no life form had any use for oxygen; it was simply a waste product. But it may have been a destructive waste product. Probably certain primitive forms of animal life had already come into being. These creatures had evolved in an oxygen-free atmosphere; their bodily functions were adapted only to this atmosphere.

The gradual increase in oxygen must have proved disastrous to these life forms. There are several living species of bacteria which thrive only in oxygen-free surroundings. These bacteria may be relics of the first generation of living beings on earth. If such primitive animal life in fact existed (and indications are that it did), it must have succumbed to the worldwide epidemic of oxygen poisoning induced by the burgeoning plant life.

We now think of oxygen as a friendly and indispensable element. The universal oxygen poisoning of our atmosphere might have spelled an end to all life on this planet. Even the plants might eventually have suffocated from the effects of their own waste product. Instead, nature began all over again, creating animal forms adapted to an oxygen atmosphere. These creatures did not merely tolerate the new atmosphere; they made a virtue of necessity and used the oxygen to produce energy. What a striking proof of the tenacity and adaptability of life!

These remarkable new creatures solved two problems at once. We have already noted that the oxygen atmosphere threatened plant as well as animal life. Oxygen might have continued to accumulate indefinitely, eventually suffocating the plants in their own wastes. The newly evolving creatures prevented this from occurring; they consumed the waste material of the plants. The carbon cycle came into being, establishing a biological equilibrium. Oxygen ceased to accumulate in the atmosphere and leveled off at around twenty-one percent.

It was fortunate for the plants that the oxygen level rose no higher than this. Recently some interesting experiments have been conducted on plants. The plants were raised in artificial environments containing a wide variety of chemical compounds. Scientists wished to discover whether earth plants could survive in an alien atmosphere like that of Venus or Mars. They were amazed to learn that the present composition of the earth's atmosphere does not constitute the ideal environment for plants. All the plants thrived in an atmosphere containing only half as much oxygen as the earth. They grew to almost double their normal height and developed more luxuriant foliage than ever before.

We tend to view our environment in terms of our own needs. For centuries we believed that the earth was the center of the universe—simply because it was the center of *our* universe. In the same way, we look on plants as our servants; they are useful organisms which supply us with oxygen and food. The historical development of the earth's atmosphere shows us the fallacy of this view. Plants are not the servants of men and animals. It might be truer to say that men are the servants of plants. From a plant's point of view, animals and men represent helpful and efficient life forms. We animals took over the task of disposing of plant waste which would otherwise have smothered all plant life.

The carbon cycle renews the oxygen in our atmosphere. Like a spaceship, our earth has a limited oxygen supply. This original supply could last only three hundred years. But for millions of years, our ancestors have been using the same oxygen again and again. Scientists plan to incorporate this recycling principle into future manned spaceships designed for long voyages in space.

We use the same air over and over. This also holds true of the water we drink. Since the dawn of time, countless generations of plants, animals, and men have drunk the same water time and time again. All living organisms require water, which acts as a medium for chemical or metabolic processes taking place in the body. And all living organisms eliminate this water again. The kidneys use it as a solvent in which to eliminate poisonous waste products. Perspiration eliminates some water, helping to regulate body temperature. We also exhale a certain amount of moisture. As we exhale, this moisture keeps sensitive mucous membranes from drying out–mucous membranes located in the respiratory tract, from the finest branches of the bronchial tubes to the skin inside the nose.

Like air, water passes through a cycle of renewal. We are all familiar with this cycle. The used water finds its way to rivers or streams and then to the sea, where it evaporates in the sun's heat. In the atmosphere the vapor condenses into clouds composed of countless tiny drops of pure distilled water. Heat from the sun stirs up air currents that transport the clouds to other regions. Eventually rain restores the purified water to the land.

And now, what about food? The Spaceship Earth has a limited food supply. How can it provide for its ever-renewed crew of animals and men? As schoolchildren, we learned how food is created, but we were not taught to look at the whole picture. Before we can fill in this picture, we need a little more information.

To understand what food really is, we must first understand the basic rule, "Nothing comes from nothing." Scientists call this rule the law of Conservation of Energy. This law applies in the inorganic world of physics; but it also applies to biological processes. After all, biological processes also consume energy. More precisely, any process involves the conversion of certain forms of energy into different forms of energy. At this point the famous law of entropy comes into play. The conversion of energy follows a fixed pattern. A certain portion of the converted energy is transformed into heat. This heat becomes uniformly diffused in the atmosphere, thus ceasing to function as a possible energy source. The energy released as heat appears to have been lost. In reality this phenomenon does not contradict the law of Conservation of Energy; the energy is not lost, but simply converted into another form of energy.

All living things require a constant source of energy. They consume energy in their metabolic processes and in physical exertion. In fact, almost all creatures need energy simply to carry their own weight. Living creatures absorb energy from many sources. Human beings absorb some radiant energy from the sun. We all know from experience that our appetite seems less keen on hot summer days than during the winter. Sunlight directly supplies some of our energy needs.

But sunlight can never supply us with all the energy we need. We cannot satisfy hunger merely by sitting in the hot sun for a few hours. Animal organisms require an additional energy source. In other words, animals and men expend energy more rapidly than they can absorb it from sunlight. There is another reason why sunlight does not satisfy hunger: It does not supply the body with building materials. Food supplies both energy and raw material. We use organic material to renew body tissue—muscles, fatty tissue, blood, bones, and so forth. The different types of tissue grow at different rates. Our food consists primarily of fats, proteins, and carbohydrates. Our bodies also require other substances: vitamins, minerals, and trace elements. None of these supply energy; they are needed as building material. Without them, we would suffer from various forms of malnutrition or vitamin deficiency.

The fats, proteins, and carbohydrates in our food that supply energy and organic building material are composed of molecules of high molecular weight; that is, they are formed from complex molecules made up of many individual atoms. These foods supply energy because of their high molecular weight. Complex molecules use energy to bind their many atoms together; the food in turn transmits this "binding" energy to us. Our bodies break down the molecules into simpler compounds of lower molecular weight. During this process, some of the energy contained in the original molecules is released into the body. Some of the energy takes the form of heat, which maintains our body temperature. But some of it also appears as electrical energy or free electrons. No doubt it takes other forms as well. We do not clearly know how the body makes use of these various forms of energy.

Let us return to our original question: What is the source of our food? To put it a little more clearly: What is the source of the energy stored in certain compounds like proteins or sugars? Where did it come from, and how did it get into these compounds? We have already noted that the basic principle, "Nothing comes from nothing," applies to both organic and inorganic processes. The energy in our food must have come from somewhere.

We have mentioned that plants use the sun's energy to build organic compounds. The sun provides us with light and heat. But indirectly it also supplies us with food. In the past, people often spoke of the sun as the "life-giving star." Today this description seems more fitting than ever. In recent years we have learned how profoundly the sun has influenced life on earth. We shall return to this subject in later chapters.

Indirectly, then, it is the sun which feeds us. We know that the sun's rays cannot satisfy hunger directly. Food reaches us through a "food chain" which originates in the sun. The green leaf pigment of plants, known as chlorophyll, absorbs certain wavelengths of the sun's radiation. Plants also absorb carbon dioxide from the air; their roots take in simple inorganic molecules from the soil. Then plants combine all these elements into various compounds of high molecular weight. Thus it is the leaves which actually absorb the sun's energy, using it to bind simpler molecules into organic compounds. Among these compounds are the carbohydrates, fats, and proteins. Plant leaves supply all the food energy in our world. Without plants to supply us with oxygen, we would suffocate sooner or later. But probably we would starve to death long before our oxygen ran out. The meat on our table comes from animals which fed on plants or on plant-eating animals. The food chain always leads back to the plant kingdom.

At the beginning of this book, we encountered an unusual example of the food chain. Living creatures inhabit even the deepest levels of the ocean, more than six miles beneath the surface. The water pressure at this level amounts to one metric ton per square centimeter. Until recently scientists believed that few living organisms could survive such pressure. Plants do not exist in this environment. No ray of sunlight can penetrate such depths; therefore plants have no means of nourishment. Nevertheless, many forms of animal life inhabit this region. They include various fish and eels (many known only by their scientific names), as well as sea cucumbers, sea anemones, sponges, crabs, and sea lice.

What do these deep-sea creatures do for food? Here as elsewhere, no doubt the large eat the small. It may appear that these animals survive quite independently of plants. But the chemical analysis of deep-sea samples proves the contrary. The food chain even extends to the depths of the sea. Dead animal and plant remains filter steadily down to the ocean floor. The organic compounds they contain nourish the creatures of the deep, which thus live on the refuse of a world they know nothing about. The world above provides them with food; yet they would find this life-giving world as alien and hostile as we would find the surface of the moon.

The Earth Is Not an Autarky

OUR Spaceship Earth carries a cargo on its long journey through space, a cargo that includes everything needed by the crew—oxygen, water, and food. Only one thing is missing from the cargo, the energy needed to replenish these supplies. This energy must come from outside the earth. We have described the cycles which renew our oxygen, water, and food. Great quantities of energy are needed to maintain these cycles. Each year, between 600 and 700 trillion metric tons of water evaporate from the equatorial regions, rising into the atmosphere. Air currents carry them north and south toward the poles. Soon the fresh water rains down on the earth. Great heat is required to evaporate the water and transport the clouds. The energy used in this process comes from the sun. The sun also supplies the radiant energy used in photosynthesis—the process that provides us with oxygen and food. Utilizing the sun's radiation, plants produce more than 200 billion metric tons of organic matter each year!

The sun is the ultimate source of all energy. It acts as the "nuclear propulsion unit" of our Spaceship Earth. This propulsion unit does not actually propel our planet through space; but it does function as the driving force behind the life processes taking place on earth. The sun releases its energy in the form of electromagnetic waves which supply us with heat and light. This radiation travels 93,000,000 miles to reach the earth. Plants act as antennae to receive the radiant energy needed to support life.

Generations of scientists have tried to understand the huge fire burning in the sky overhead. We all lead our lives by the light of this fire. For centuries men have watched it burn, and yet it has never burned out. Until recently, scientists could not discover the source of the vast amounts of energy emitted by the sun.

Let us try to picture the sun a little more clearly. It lies almost 93,000,000 miles away. (We have already tried to picture this distance by using a scale model: the grapefruit-earth orbiting a sun one mile away.) Another example may help us to understand the distance separating the sun from the earth. If sounds could travel through space at the same velocity as they travel on the earth, then sounds on the sun would have to travel 14½ years to reach our planet. If an explosion took place on the sun, we would not hear it for 14½ years. (This is a purely hypothetical example. The earth and the sun are separated by 93,000,000 miles of airless space. No sound could ever traverse that space, since there is no medium to carry it.) Nevertheless, this distant star burns so fiercely that its light can pierce heavy clouds above the earth; and it drives us into the shade on a hot summer day.

Hot as we may feel during the summer, we should keep in mind that the earth receives only $2/1{,}000{,}000{,}000$ of the total energy emitted by the sun in any given moment! We lie at an immense distance from our sun; moreover, the surface of our planet offers a very small target. These two factors explain why we absorb such a tiny percentage of the sun's energy. Only ten times this amount—$2/100,000,000$ of the sun's total radiation—is absorbed by all the planets in our solar system put together. The rest of this energy streams out into the depths of space, where it all appears to be lost—one more example of nature's occasional senseless extravagance. Until recently, everyone assumed that this energy served no purpose. We now know that some of the radiation freely diffused into space plays an important role in life on earth. In fact, our lives depend on this "wasted" energy; we need it as much as we need the sunlight that actually reaches the earth. A later chapter will discuss this energy and how it was discovered.

But first let us return to our original question: What makes the sun burn so brightly, yet never lets it burn out? We now know that our sun is an atomic oven. Until about forty-five years ago, scientists did not understand the principle of atomic energy. Therefore they were baffled by the sun's behavior.

In his *Universal Natural History and Theory of the Heavens*, already mentioned in the chapter called "Astronautics and the Dimensions of Space," Kant gives a dramatic account of our sun. He describes the sun's surface as a giant sea of fire fed by molten masses of combustible material erupting from the sun's cold interior. Ever since Galileo, men had been aware of sunspots. Kant interpreted the sunspots as tall mountain peaks which now and then briefly emerged from beneath the glowing flood.

Even now, many people have a false impression about sunspots. Sunspots look to us like black flecks standing out against the white-hot disk of the sun (Illustration 13). We can observe them in a variety of ways: directly through tinted glass or a special telescope; or indirectly on a photograph or on a screen linked to a telescope. No matter how we view them, they always look black on white. Inevitably, we imagine that they must be cold areas of the sun's surface. Prominent scientists have had the same experience. Kant was misled by their appearance, as was the famous Sir John Herschel (1792–1871). Herschel believed that the sun's surface was cool, and that only its outer atmosphere was actually burning. He interpreted sunspots as patches of this cool surface visible behind the burning atmosphere. Later scientists surmised that the spots were areas of sediment or cinders. Indeed, the spots really *are* 1,500° Celsius cooler than the surrounding area, which appears white by comparison. Fifteen hundred degrees is nothing to sneeze at. But the normal temperature of the sun's surface is 5,700°C. Thus the spots have a temperature of 4,200°—far hotter than white-hot steel!

Let us suppose that we could remove one of these dark-looking spots from the sun and place it in the sky somewhere about 93,000,000 miles away. This spot would be no larger than one of our planets, like the evening star. It would no longer appear dark. In fact, at night it would light up the earth as brightly as a full moon! Nevertheless, sunspots continue to look black in photographs. This effect results from our photographic techniques. The sun's brilliance might easily block out the image of the spots altogether. In order to see the spots, we must first darken the sun's light by means of strong filters. Inevitably, these filters also darken the sunspots. Many amateur photographers have faced a similar problem. Suppose a person wants to photograph the sunlit façade of a white house. He wishes to include the open front door in the picture, but the open door reveals a shadowy hall. He now has two choices. He may do nothing to dim the bright façade; then the door will scarcely show up in the picture at all. On the other hand, he may choose to tone down the façade. But now the door ends up looking like a black hole. In reality, the hall is bright enough to read in. (See Illustration 14.)

Kant and later scientists assumed that the fire on the sun resembled fires on earth, that is, the fire fed on oxygen and consumed combustible material. At that time no one knew how far the sun's rays had to travel to reach the earth. If scientists had realized how far away the sun really was, they would have seen at once that no ordinary firelight could have traveled the 93,000,000 miles to our planet. There was another flaw in their theory: If the sun's fire resembled fires on earth, the sun would have burnt out long ago. Suppose the sun really consisted of combustible material. It might be a huge ball of high-grade bituminous coal, almost 1,000,000 miles in diameter (the diameter of the sun amounts to four times the distance from the earth to the moon; if we hollowed out one half of the sun, we could place the earth inside it—still being orbited by the moon!). Even such a giant mass of coal would burn out in a mere 25,000 years. Kant and his contemporaries did not know the true size of the sun. Nevertheless, they realized that something was wrong with their theory. How could the sun have gone on burning so long? They tried to explain this apparent flaw in their argument by suggesting that the sun might be continually acquiring new fuel from passing comets and meteors. In Kant's time, scientists believed the earth to be no more than 100,000 years old. If the sun had acquired new fuel from passing comets, it might easily have survived the 100,000 years of the earth's existence.

For a time this theory seemed to fit the facts—but not for long. The youthful science of paleontology at first appeared quite unrelated to astronomy. Yet it was the paleontologists who threw a monkey wrench into the works. They had been systematically scouring the earth's crust in search of fossils, the petrified remains of now-extinct life forms. With ever-increasing accuracy they were able to determine how much time had passed since the death of these fossilized creatures. Paleontologists kept finding fossils far more ancient than the supposed 100,000 years of the earth's existence. (See Illustration 15.)

By the turn of our century, scientists were estimating the age of the earth in hundreds of millions of years. Paleontologists now possessed data proving the presence of organic life on earth almost 200,000,000 years ago. Thus the earth itself must have been even older. Moreover, life on earth could not exist without energy from the sun. Therefore the sun must have been burning throughout the entire 200,000,000 years. Once more, the astronomers had their backs against the wall. They tried to maintain their theory that meteorites had continually replenished the sun's fuel supply. But they soon saw that they had reached a dead end. If the sun had been attracting large quantities of cosmic debris, this material would have caused fluctuations in the sun's weight. But astronomers had already "weighed" the sun. They had used the most precise scale in the world: the planetary orbits. If the sun had periodically increased in mass, then its gravitational pull would have shifted the orbits of the planets, but astronomers knew that these orbits had been stable for hundreds of years.

Then astronomers pulled a new rabbit out of the hat. They suggested that the sun's own gravitational field was causing it to gradually contract. The sun weighs more than 330,000 earths; therefore it has an immensely powerful gravitational field. Moreover, the sun originally formed from contracting gas. Thus it seemed entirely possible that the sun was still contracting. The contraction of any body generates heat. If the sun were contracting, it might produce enough heat to maintain its own burning. (We now know much more about this process of contraction. Hydrogen contracts in giving birth to a star. This contraction creates heat, which sets off nuclear reactions in the star. Nuclear energy keeps the star alive.) Astronomers made the following calculation: If the contracting sun lost only 1/10,000 of its diameter every thousand years, it could maintain its temperature at the same level. They believed that this process could have continued for several million years. Once again, they thought they had found a possible solution.

And once again, it was the paleontologists who upset the applecart. By the 1920s, paleontologists agreed that life had existed on earth for nearly one billion years. The contraction process could not have continued to maintain the sun's temperature for this long a span. The astronomers began to rack their brains once more. The answer came in 1925, when Sir Arthur Stanley Eddington suggested that atomic energy created the radiation of stars. Scientists were already familiar with the theory of atomic energy, but Eddington was the first to relate this theory to the energy production of the sun.

Paleontologists now tell us that life has existed on earth for around 3 billion years. Atomic energy could easily have sustained the life of our sun for such a period. In fact, it appears that our sun has been shining with undiminished brilliance for some 4½ billion years. That is, it has already lived the first half of its life. Of course, the sun will continue to exist even after another 4½ billion years have gone by. But it will no longer be the sun we know. It will undergo a series of crises, during which our earth will be destroyed along with the entire solar system.

Thus an atomic oven warms and illuminates our world; this oven maintains our life cycles. But our "life-giving star" does far more for us than this. Recently scientists have discovered a series of hitherto unsuspected ties between our sun and living conditions on earth. To understand these ties, we must first understand the huge atomic reactor that gives us life. We must try to draw the portrait of a star. Everything about this star seems awesome and enigmatic. Much of the inconceivably hot furnace at its core would appear dark to human eyes. And the light it pours through our windows was born in the Stone Age!

Portrait of a Star

THE story of our sun began between 6 and 8 billion years ago when a gigantic cloud of interstellar matter gradually began to contract. This finely diffused cloud was composed of hydrogen atoms with a small admixture of heavier elements; initially it was many hundreds of times larger than our present solar system. The contraction of the cloud resulted from the mutual gravitational attraction of the hydrogen atoms. As the atoms contracted toward their common center of gravity, the center of the cloud continually grew more dense. The increasingly massive cloud core gradually intensified the gravitational pull it exerted on outlying atoms. As the core increased its attraction, the cloud contracted faster and faster. This process continued over an immense span of time. As they contracted toward their common center of gravity, the hydrogen atoms tended to graze or bump into one another. From the very beginning, the cloud had been undergoing a second form of motion; while continuing to shrink, the entire formation was also turning on its axis like a giant carousel.

As the contracting cloud grew smaller, it began to spin faster and faster. That is, its increasing compression also increased its speed of rotation. The rotation created centrifugal force, which tended to flatten out the formerly spherical cloud. Eventually a fragment of the now disklike cloud broke away. Although this shred comprised less than one percent of the cloud's total mass, it carried away about ninety-nine percent of its impulse of rotation. As its rotation slowed, the disk began to resume the shape of a sphere. In time the shred of matter which had broken away from the contracting cloud formed the nine planets. The planets retained the impulse of rotation that had been torn away from the cloud. Throughout the time the cosmic carousel had been turning, the plane and direction of its rotation were permanently fixed, for no outside force was powerful enough to shift the carousel's orientation in space. Since the cloud had already established the plane and direction of its rotation, the planets had to follow suit. Today all our planets lie in the same plane and revolve in the same direction.

The present structure of our solar system helped scientists to reconstruct its earlier phases of development. That is, the present-day position of our planets gave astronomers clues to the way in which the sun and the solar system must have formed. We know that certain things must have occurred in the past to produce the phenomena we observe in the present. To be sure, this rule holds true only if we make one basic assumption: that throughout the history of the universe the laws of nature have not altered. (As already mentioned in the chapter called "Astronautics and the Dimensions of Space," perhaps we may one day learn that these laws *have* changed.)

We have noted that as the cloud particles continued to contract toward the core, the core increased in density and gradually intensified its gravitational attraction. After the cloud fragment had broken away from the disk, the cloud's powerful gravitational field overcame the remains of the centrifugal force which had tended to flatten the formation. That is, the attraction of the core caused the cloud to resume the shape of a sphere. This sphere became our sun. Almost all of the original cloud went into the formation of the sun, which now accounts for almost 99.9 percent of the total mass of our solar system. Slightly more than .1 percent of the total mass makes up all the planets, meteors, comets, and interstellar dust. In terms of sheer mass, the sun is not merely the center of our solar system, it *is* our solar system. The nine planets (including our earth) and the rest of the system weigh virtually nothing in comparison.

As the huge cloud of matter continued to contract, it unleashed mysterious processes which have never existed on earth. For all their achievements, our modern technicians are unable to reproduce these processes under laboratory conditions. Somehow a mere ball of hydrogen began to evolve into a true sun; it developed properties which continue to astound the scientists who discover them. Most amazingly of all, these special qualities of our sun made possible the evolution of life on earth. Remote and alien as this star may seem to us, it is both the source and the guarantor of our existence.

As the giant sphere contracted, it steadily grew more compressed. This growing density built up enormous temperatures and pressures at the core. At last the tremendous heat and pressure set off nuclear fusion reactions in the core. These atomic reactions continued to release huge quantities of energy. With the development of atomic energy, the gaseous sphere became a genuine star. At the same time, the newborn sun entered into a crucial phase of "instability."

As long as the gaseous cloud was still contracting, it clearly could not be considered stable. On the other hand, for untold aeons the cloud had contracted at a fairly uniform rate, picking up speed very slowly. Moreover, no outside force had interrupted the process of contraction. Thus the contracting cloud had maintained a state of relative stability. But the onset of atomic reactions in the core altered this situation. The atomic reactions created energy which pressed *outward*, slowing down or even reversing the process of contraction. That is, atomic energy counteracted the gravitational attraction of the star's core. We do not know exactly what happened next; no traces of the sun's early past are visible now. But presumably the sun responded to the sudden increase in internal pressure by temporarily beginning to expand again.

A strange chain of events ensued. As the sun expanded, its internal pressure and heat began to subside. In other words, the pressure and heat fell below the critical levels needed to produce atomic reactions. The newly kindled atomic fire went out. Atomic energy ceased to push outward; thus it ceased to counteract the gravitational attraction of the core. The gases began to contract once again. The contraction built up the heat and pressure in the core until they induced new atomic reactions. Atomic energy forced the sun to expand once more, thus keeping the cycle going. We do not know precisely what occurred; probably the sun passed through a period of fluctuation. During this phase it must have expanded and contracted in a series of rhythmic pulsations.

During this phase of the sun's evolution, life could never have developed on the earth. Of course, at this time the earth may not even have existed. If it did exist, the sun's pulsations would have caused radical temperature variations on our planet. Whenever the sun was in its contracted state, temperatures on earth would have been very similar to what they are today. But during the sun's periods of expansion, the heat on the earth would have precluded the existence of life as we know it. This statement may sound paradoxical. After all, when the sun was expanding, its gases would have grown cooler. At the same time, this expansion would have halted the heat-producing nuclear fusion process in the sun's core, further reducing the sun's temperature. How could a cooler sun have produced a hotter earth? The solution to the paradox is this: As it expanded, the sun's surface would have come much nearer the earth. Although the sun had grown cooler, its surface would still have been scorching hot. It would have driven temperatures on earth up to several hundred degrees Celsius.

We do not know how long our sun continued its rhythmic pulsations. In terms of astronomical time, it cannot have been for very long. If newborn stars continued to pulsate for a prolonged period, we would be able to see many such pulsating stars in the sky today. After all, new stars are being born all the time. These new stars must all pass through a period of rhythmic pulsation. We could not actually *see* the pulsations occurring—not even through the most powerful telescope, for even the nearest stars are too far away for us to view their movements from the earth. But astronomers do know of many stars which frequently grow brighter or dimmer for periods of days or even weeks. Indications are that these stars undergo a rhythmic pulsation. Some of them may well be "young suns" in the very first stage of their existence as stars.

In any case, we know that eventually our sun became stabilized. The sun has remained stable for well over 4 billion years; probably it will maintain its equilibrium for at least another 4 billion years. In time, the young sun's internal pulsations grew weaker and weaker. At last the pressure of expansion exactly equaled the pressure of contraction. The sun's present state of equilibrium guarantees the equilibrium of our solar system. If this equilibrium were even slightly disturbed, the entire solar system would suffer instant annihilation. It is comforting to realize how long the sun has remained stable and how long it is likely to do so in the future. On the other hand, even a star is mortal—as we shall soon see.

Our sun is a star in a comparatively stable phase of its existence. All stars pass through a succession of phases—an entire life cycle. As human beings, we are principally concerned with the "contemporary" history of our sun—the history which includes our earth. We want to become acquainted with the power that sustains all life cycles on our planet.

Let us now try to draw a portrait of the structure and appearance of our sun. The first question that arises is this: How do we know what goes on in our sun, or how its interior is constructed? No one can see inside the sun. Only two areas of the sun can be directly observed by our instruments, the outer atmosphere, and portions of the scorching surface just beneath the atmosphere. One day we may be able to see more. Scientists are already discussing the possibility of a so-called "neutrino astronomy."

Neutrinos are basic particles, parts of an atom, with certain unusual properties. One might almost say that their most unusual property is their lack of any properties whatsoever. A neutrino has no electrical charge and virtually no mass; it has only a torque, a so-called "moment of spin." Imitating Morgenstern's famous poem,* a physicist once said of the neutrino, "It's a spin, that's all it is." This virtual lack of properties may prove a useful trait. A particle without charge or mass meets with no resistance when encountering a solid body. Of course, even a so-called "solid" body is simply a swarm of atoms; it contains far more empty space than impenetrable matter.

And yet if one solid body comes in contact with another solid body, it generally encounters resistance. If I strike my fist on the table, the resulting crash derives from the fact that the atoms of both objects are tightly bound together and therefore cannot pass through one another. A bird is unable to fly through a chicken-wire fence, which appears to be mostly air, because the wires that make up the fence are attached firmly together. On the other hand, a mosquito is able to pass easily through the same chicken-wire fence because it is much smaller than the space

* The reference is to a humorous line of twentieth-century German poet Christian Morgenstern (1871–1914): "It's a knee, that's all it is." The knee in question goes walking through the world all alone, unaccompanied by a body. Thus it resembles the "bodiless" neutrino (translator's note).

between the wires. A neutrino is something like a mosquito for which all ordinary matter is a chicken-wire fence.

The nuclear fusion process in the sun's core is constantly creating more neutrinos. These neutrinos pass right through the sun at the speed of light; if they come our way, they nonchalantly fly straight through the earth. We are all constantly being peppered with a whole stream of neutrinos; yet we never notice them and are not harmed by them. All these neutrinos originate in the sun's core. If we were able to capture and investigate them, we could learn more about the processes taking place inside the sun; it would almost be possible to "see inside" it. A few successful attempts have been made to trace neutrinos. One day, as we have stated, there may really be a neutrino astronomy. But at the moment this idea is still just a castle in the air. Since neutrinos pass almost effortlessly through all known forms of matter, we would first need to develop an instrument capable of holding on to them long enough for us to observe them. We have not yet worked out even a theoretical solution to this baffling technical problem.

Yet we do know with astonishing precision just what the interior of the sun is like and what is taking place there. We owe this information to an invention which few people would think of associating with astronomy, the computer. There is a good reason why computers are used to investigate our sun.

Astronomers found it a fairly easy task to determine many of the sun's basic characteristics: its size, weight, distance from the earth, and the quantities of energy it radiates each moment. The sun's diameter is 864,000 miles; it weighs 330,000 times as much as the earth; its mean distance from the earth is 92,900,000 miles; and each square centimeter of its surface radiates 1,500 calories of energy per second. Astronomers have also subjected sunlight to spectroscopic analysis in order to determine its chemical composition. About seventy percent of our sun consists of the lightest and simplest of all elements, hydrogen gas; the remaining thirty percent consists almost entirely of the second lightest element, helium; two or three percent at most is composed of heavier trace elements.

After making these relatively easy calculations, astronomers possessed basic data regarding the sun. They could use these facts as stepping stones to new information. They had only to reconstruct what was going on inside the sun that could have produced the results they saw on the surface. But the physical facts at their disposal were so numerous, the interrelationships so complex that the task seemed endless. An entire company of patient and dedicated mathematicians could hardly have carried out the necessary calculations. Moreover, they would have had to perform all these complex computations many times over. First astronomers would have needed to set a hypothetical case for conditions in the sun's interior; then they could have performed all their calculations in order to find out whether the hypothetical case might have produced the known conditions on the sun's surface. This would have been a hit-and-miss procedure, a true labor of Sisyphus. Astronomers would have had to go on constructing and discarding theoretical models until they had found one that fit. Only modern electronic computers were truly capable of carrying out this laborious and time-consuming experiment.

Thanks to these computers, we now know how the sun looks inside and what processes are taking place within it. The story of our sun is a catalog of the astounding and the improbable. Our imaginations are geared to the sights and sounds of earth. For us, conditions in the sun are in every sense out of this world.

The pressure at the sun's core equals more than 200 billion metric tons per square inch. The temperature reaches 15,000,000° Celsius. At this pressure, the matter at the sun's core is twelve times as heavy as lead—and yet it still takes the form of a fluid (plasma). At such temperatures the hydrogen and helium atoms are totally ionized; they have lost their electrons and consist of nuclei alone. We have already noted that the major portion of an atom is simply "empty space"; the distance between the atomic nucleus and its orbiting electrons greatly exceeds the size of the particles themselves. But the nuclei stripped of their electrons can be packed together to an abnormal degree of density. They form a special kind of matter known as "plasma." Despite its great density, this matter at the sun's core retains the properties of a fluid—above all, the ability to circulate freely, creating areas of free flow or turbulence.

Exactly how hot is 15,000,000°C? Again we are confronted with a number our minds cannot grasp. But the famous English astronomer Sir James Jeans (1877–1946) once computed the effect such a temperature would have on our everyday surroundings. According to Jeans, if we removed a bit of matter *the size of a pinhead* from the sun's core and then placed it on the earth, its heat would kill a man ninety-four miles away!

This temperature generates atomic reactions in the ultradense matter of the sun's core. These atomic reactions release the energy which the sun so lavishly squanders in all directions. We must take a closer look at the process of nuclear fusion.

At one time or another we have all heard that the sun's energy is produced by the fusion of hydrogen into helium. This fusion process takes place in several steps. First, about once in 7 billion years each hydrogen nucleus in the sun's core crashes head on into another hydrogen nucleus, thus producing one helium atom. Despite the abnormal density of the sun's core, and despite the internal flow induced by the 15,000,000° temperatures, the tiny hydrogen nuclei still have lots of space to move about in. This explains why a direct collision between two hydrogen nuclei is such a rare occurrence. The "active" zone of the sun—the region in which atomic reactions occur—has a diameter of about 218,000 miles, a distance roughly corresponding to the distance between the earth and the moon. This enormous zone contains vast quantities of hydrogen nuclei. Even if each of them joins in a fusion reaction only once in 7 billion years, there are always enough nuclei to keep the process going.

But the supply of hydrogen atoms will not last forever. In the fusion process, four hydrogen nuclei are fused to produce one helium atom. Each second, 657,000,000 metric tons of hydrogen are converted into 652,500,000 metric tons of helium. The helium may not be used for further energy production. It piles up, forming the "ash heap" of the atomic process. This explains why our sun will not continue to shine for all eternity. Present indications are that it has already been shining for between 4 and 5 billion years. It must have already consumed more than half of its hydrogen fuel.

Six hundred and fifty-seven million metric tons of hydrogen converted into 652,500,000 metric tons of helium every second for nearly 10 billion years (the sun's total life span)—these figures should give us some notion of the magnitude of our sun. But what of the difference in mass revealed by these figures? What becomes of the 4,500,000 metric tons of matter apparently lost every second? Why does the conversion of one element into another slightly heavier element produce a reduction in mass? The explanation is this: A helium atom actually weighs slightly less than the sum of four hydrogen atoms. A tiny fraction (less than one percent) of the total weight of the atoms is left over from the nuclear fusion. This fraction must be disposed of in some way. It is changed into neutrinos and pure energy which leave the sun and travel into space. Each second the energy released by the destruction of more than 4,000,000 metric tons of matter radiates from the sun. The sun produces more energy in one second than man has produced in all of human history. In three days, the sun emits as much energy as we could produce by burning all the coal, oil, wood, and other combustible materials on our earth.

The sun produces enormous quantities of radiant energy; and yet the vast bulk of this energy is absolutely invisible. If it were possible for us to spend some time in the sun's core, we would see relatively little of what is going on here, for much of the radiation being produced is of such high energy that it would simply be invisible to our senses. This phenomenon is similar to the "inaudible" whistle a hunter uses to call his dogs. The whistle emits high notes inaudible to the human ear and to wild game; but the hunting dog reacts to it at once. Although there is some visible energy at the sun's core, most of the energy consists of the most intense forms of radiation, gamma rays and X rays. Both are invisible to human eyes. As this invisible energy makes its way from the sun's core to its surface, it is "softened" into the visible light which is radiated out into space.

The sun comprises virtually our entire solar system. But its very bulk is essential to our lives. If the sun were not so large, it could not build up the enormous temperatures and pressures in its core, that is, it could not generate atomic reactions. Moreover, the sun's bulk acts as a shield to protect us from the energy at its core. If this energy were to strike the earth in its original form—gamma rays and X rays—none of us would be alive today. But because our sun is so vast, the earth and its fellow planets are bathed in light and warmth rather than in deadly radiation. Each quantum of energy has to make its way from the core of the sun to its surface; it has to cross through about 373,000 miles of solar matter. Only then is the energy "loosed" upon the planets. This journey of energy from the sun's core to its surface comprises almost twice the distance from the earth to the moon. Moreover, the energy travels not through empty space, but through abnormally dense matter. This matter is particularly dense during the first stretch of the trip.

If we examine the path the energy takes in traveling from the sun's core to its surface, we find that it literally bounces along its way. The energy quanta move at the speed of light. All the same, it takes them an astonishing length of time to reach their goal. It takes them so long because they cannot follow a direct route to the surface. Instead they must pursue an adventurous zigzag course; they are continually absorbed and then released again by all the atoms lying in their path. These intervening atoms greatly delay the energy on its journey. Computers calculate that a single quantum of energy travels 20,000 years to reach the sun's surface. Thus the sunlight that pours through our windows today was actually created in the Stone Age!

In a sense this statement is not fully accurate. The light shining through our windows is not literally the same light created in the sun's core. The original light gradually became "exhausted" during its long and laborious journey to the open sky. The light which has finally worked its way to the sun's surface arrives on earth only eight minutes later; but by now this light is the "ghost" of its former self. Yet this dim reflection of past glory provides just the right amount of light and heat to nurture the frail creatures of earth.

The Solar Wind

DURING the past decade, revelations in astronomy have been coming thick and fast. Recently astronomers have learned that the sun emits other forms of energy besides electromagnetic radiation. Like all atomic reactors, the sun generates a second form of radiation which is material, or "corpuscular," in nature. This corpuscular radiation consists of individual particles in the form of swift-moving atomic nuclei and electrons. When these particles leave the sun's surface, they are traveling at a speed of more than 310 miles per second. As they fly past the earth several days later, they are still moving at almost a thousand times the speed of sound.

The discovery of these particles has cast new light on the role the sun plays in our daily lives. Our sun is indeed the "life-giving star." It not only supplies us with energy we could acquire from no other source, but also protects us from deadly threats that arise in outer space. In order to understand how the sun protects us, we must first become more familiar with the stream of particles known as the "solar wind."

For some time scientists had been aware that a mysterious force was emanating from the sun. This force was able to "blow away" matter that came near to the sun's surface. At times astronomers could directly observe the effects of the mysterious "wind," for something special happened whenever a comet flew quite near the sun.

Comets are very striking in appearance and crop up quite abruptly in the sky. For these reasons, the "tailed stars" were once considered harbingers of war and pestilence. In reality comets are rather small cold chunks of matter with diameters of only .6 to 60 miles. Our own solar system contains many comets. These chunks of matter travel along highly eccentric elliptical paths around the sun (see Illustration 16). Their orbits extend over enormous distances. Some comets must travel for thousands of years to complete a single orbit around the sun! At some points of its orbit, a comet may be as far as two or three light-years from its sun.

The gravitational pull of the sun is so powerful that it continues to attract a comet two or three light-years away. That is, the force of gravity still binds the comet to our solar system. But such a comet is on a "long leash": two or three light-years amounts to about half the distance separating two neighboring solar systems. The distance between any two fixed stars is so enormous that most likely our astronauts will never be able to travel to even the nearest solar systems. Nevertheless, direct physical contact does occur between neighboring solar systems, for comets can sometimes do what we cannot. When a comet is farthest from its sun, its path frequently crosses the path of a comet from another solar system; this second comet is equally distant from its own sun. When the two comets cross paths, they may exchange suns. Each comet begins to orbit another, alien sun.

A second factor may contribute to this "exchange of comets." When comets travel near their suns, the influence of the planets sometimes throws them off course. Planetary influence increases the irregularity of comet orbits, thus increasing the chance that a comet may stray into a neighboring solar system.

With time, the orbits of most comets become increasingly "perturbed" by the gravitational pull of the planets. Sooner or later some planet always succeeds in capturing and destroying a comet. Most comets survive "only" about 1,000,000 years; then they crash into a planet in the form of meteors, or "shooting stars." Thus some "unearthly" matter exists on our earth, part of which came from dying comets. Some of these comets were visitors from other solar systems somewhere in our cosmic neighborhood.

In a later chapter we will take a closer look at this cosmic exchange of matter. For now, let us continue our discussion of comets in relation to the solar wind. These wandering chunks of matter are normally cold. Their distance from the earth makes them invisible to us. But sooner or later their eccentric orbits bring them close to the sun again. At this point they cease to be cold and invisible. They take on the dramatic appearance which so terrified men in past ages. As the comet approaches the sun, the sun heats the cold matter so that hot gases stream out from its solid core. Spectroscopic analysis has shown that these gases include nitrogen and carbon monoxide. The gases are ejected from the comet's core at speeds of up to 620 miles per second; they are lit up by the rays of the sun. The bright gases create the striking tail which the comet will lose as soon as it leaves the area near the sun. This tail may attain a length of from 60,000,000 to 125,000,000 miles.

Astronomers have noticed a special feature of comet tails. It would be natural to suppose that a comet simply drags its enormous tail along behind it. But the atmosphere of space offers no

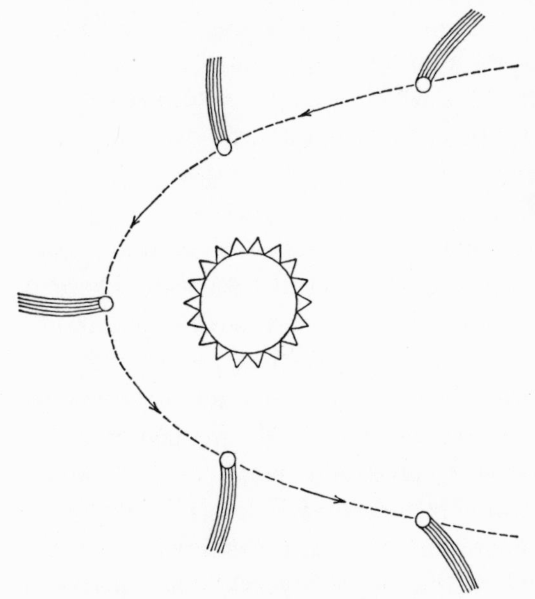

As a comet revolves around the sun, its tail always points away from the sun. This was our first indication that some mysterious force was emanating from the sun, a force that drove comet tails away from the sun's surface.

resistance, so what could force a comet to drag its tail *behind* it? Theoretically, comet tails should assume all sorts of unpredictable positions in space. But in fact this is not the case. All comets turn their tails in the same direction: *away* from the sun.

Our diagram shows what an astronomer sees when he observes a comet streaking around the sun. The diagram reproduces the positions assumed by a comet as it moves around the sun. At first the comet does appear to be dragging its tail behind it. But as the comet continues along its orbit, the tail always pulls away from the sun. In the end the comet seems to be shoving the tail before it.

The observation of comets revealed to astronomers that the sun was emitting some "repelling" force. This force caused comet tails to turn like weather vanes in the wind. Until recently scientists were baffled by this phenomenon. Some astronomers believed that the mysterious force might be the pressure of light radiating from the sun. This theory was supported by the fact that comet tails have very low density. The gas composing the tails has a density no greater than that of artificial vacuums created in laboratories here on earth. The phenomenon resembles that of the rainbow, which appears quite massive, yet actually is almost incorporeal. Gases as thin as those composing comet tails might indeed react even to the fragile pressure of light.

But other scientists maintained that the weather-vane effect observed in comets was produced by small electrically charged particles hurled from the sun into space at very high speeds. Astronomers tended to favor the "electrical particle" theory over the "light pressure" theory. They had found additional evidence that the sun was emitting electrically charged particles. Their additional evidence involved the polar, or "northern," lights (which are also found at the *South* Pole). This enigmatic play of light and color appears at irregular intervals near the polar regions; it is visible in the upper layers of the earth's atmosphere, at altitudes of fifty miles and more.

Clearly the polar lights were somehow related to the earth's magnetic field. Scientists believed that the lights were produced by electrically charged particles coming in contact with the earth's atmosphere. These particles had to be electrically charged. Why else would they appear only at the earth's magnetic poles? Moreover, scientists believed that the electrical particles were coming from the sun. A few days after increased activity had been observed on the sun, the polar lights always became especially distinct and brilliant. This "increased activity" on the sun took the form of fiery protuberances or eruptions from the sun's surface (see Illustration 17). As early as 1896 the Norwegian physicist Olaf Birkeland had theorized that the northern lights were produced by "corpuscular radiation" or some sort of "wind" from the sun. This "wind," he thought, must consist of tiny electrically charged particles. But for years there was no way to prove Birkeland's thesis.

A breakthrough occurred sixty years later, when Russia and the United States began their experiments in space. After Sputniks I and II, a third artificial satellite was launched on February 1, 1958, the American Explorer I. Explorer transmitted some startling data to its creators back on earth. Its instruments were not properly equipped to register such data, and at first scientists were simply confused by the figures they received. But later satellites relayed more precise information. The new data revolutionized our image of the space surrounding our earth. At last Olaf Birkeland's theory came into its own.

The American physicist Van Allen had suggested that Explorer I be equipped with a Geiger counter to record the presence of electrically charged particles in the earth's upper atmosphere. No one knows exactly why the other scientists complied with Van Allen's suggestion. Explorer I weighed about thirty-one pounds; it could carry only a very limited cargo. The space experts heatedly debated what hardware should be carried on the first American space mission. In any case, Van Allen's Geiger counter was part of the cargo on the day of the launch. Its inclusion brought worldwide fame to the hitherto unknown physicist.

Ironically, this instrument transmitted nothing of value. Before the flight, the Geiger counter had been working perfectly; but 600 miles above the earth, the instrument simply broke down. Very much to Van Allen's credit, he guessed the true cause of the breakdown. He theorized that the quantity of electrical particles in this region might be far greater than had been supposed. Perhaps the equipment had gone on strike simply because it had been "overfed."

To test Van Allen's theory, another less delicately calibrated instrument was launched with Explorer III just eight weeks later. This Geiger counter verified that a zone of intense radiation did in fact begin at a height of 600 miles or so above the earth. Subsequent satellite missions added to this information. It appeared that the high-radiation zone formed a belt around the earth's equatorial region. The radiation was most intense around 3,000 miles above the earth's surface. Beyond this point, the radiation temporarily decreased in intensity. At an altitude of about 12,500 miles, the satellites recorded a second radiation belt. This belt proved to be far more extensive: It extended around the entire earth, except for two rather small gaps over the poles. The gaps proved to be virtually radiation-free. Both of the radiation belts now bear the name of the man who was inspired enough to place a Geiger counter aboard Explorer I.

The discovery of the Van Allen radiation belts raised a number of disturbing questions. One thing seemed clear at the outset: Astronautical science had reached a dead end before it had even gotten started. The radiation levels reported by Explorer III and its successors would prove fatal to men in space. True, space research might continue by means of unmanned rockets and space probes. But the dream of manned space flight would never become a reality. Only lead shields weighing several tons could have protected men in space from such intense radiation. But such shields weighed too much to be incorporated into a spaceship. As far as space flight was concerned, mankind seemed to have been suddenly placed under quarantine. We had been shut in by an alien force of nature.

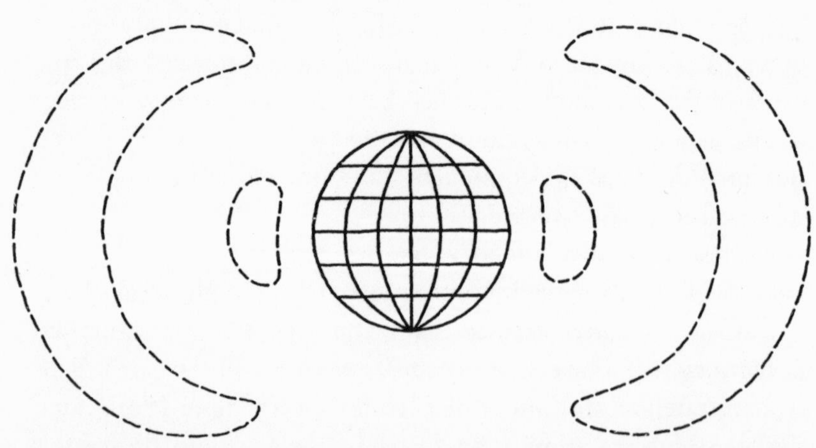

Location and extent of the two radiation belts surrounding our earth. The radiation in both belts is so intense that to spend any length of time there would result in certain death.

Scientists did not know whether they could ever develop the technological means to defeat this force. Only one solution seemed feasible: Future rocketry bases and launching sites could be built in the eternal Arctic ice. A rocket launched from one of the poles might be able to circumvent both radiation belts. Of course such a venture would have proved enormously expensive, so expensive that it might have delayed the future development of astronautics for several decades.

Fortunately, all this pessimism turned out to be premature. The radiation of the Van Allen belts is indeed deadly, but it proves fatal only when an organism is exposed to it for a considerable length of time. Moreover, the radiation of the upper belt rapidly decreases in intensity above a height of 18,000 miles. Spaceships travel at great speed. Thus astronauts are only briefly exposed to radiation; their health is not impaired. Therefore the Van Allen belts pose no real threat to manned space flight.

The discovery of the radiation belts gave birth to a new branch of science. "Interplanetary space research" involves the study of space within our own solar system. There are several reasons why interplanetary space science developed into a separate discipline. Our explorations will be restricted to our own solar system well into the distant future. Not only our instruments on earth, but even our space probes and manned space flights can investigate only a very limited area. But this is a merely external factor in our investigation of interplanetary space. More importantly, "interplanetary" space differs greatly from the "outer" space beyond it, and the difference profoundly affects our lives. A third factor has also mobilized interest in interplanetary space research: we have recently altered our basic image of space. Let us see in what ways this image has changed.

One leitmotif of this book has been man's tendency to believe that space is nothing but space—a vast emptiness dotted here and there with lonely heavenly bodies. Even scientists have tended to think this way. If an astronomer had been asked ten years ago why a man could not survive in space without technical aids, he would first have pointed out what was *absent* from space: breathable air, warmth, atmospheric pressure, etc. An astronomer answering the same question today would no doubt mention various dangerous factors which were *present:* solar "flares," cosmic rays, solar plasma, even the radiation belts.

In other words, until recently even scientists felt that space was nothing but "nothing." But data gathered by the first Explorer and Lunik flights have compelled astronomers to revise their picture of space. Space is anything but "empty." It is filled with powerful forces whose nature we have hardly begun to discover. Under normal conditions we are unaware of these forces; yet they are vital factors in our lives.

Ten years ago the discovery of the radiation belts began to transform our image of space. Investigation revealed that both belts were composed of concentrated electrical particles. The wide upper belt consisted primarily of electrons, the lower belt of protons. Small quantities of helium nuclei accompanied these protons and electrons. But where did these energetic atomic particles come from? Their only possible source was the sun. It remained to be discovered how the particles found their way into the upper layers of our atmosphere. For decades scientists had been discussing the hypothetical "solar wind." Data relayed by the Russian moon probes Lunik I and Lunik II, along with the American satellites Mariner II and Explorer X, finally verified the existence of the "solar wind" that physicists had wondered about ever since the days of Olaf Birkeland.

The sun emits huge quantities of electromagnetic radiation, above all, light and heat. But it also emits corpuscular radiation in the form of protons and electrons which fly from the sun at more than a thousand times the speed of sound. These particles are hurled straight out from the sun's surface. But the sun's swift rotation creates a sort of "lawn sprinkler" effect: It forces the particles to follow long spiral paths. (The sun turns once on its axis every twenty-five days, an extraordinarily high speed of rotation for such a huge sphere.)

The sun emits nonmaterial radiation in the form of electromagnetic waves. But it also streams out physically and materially. It "bleeds" into space. Satellite data indicate that each second the sun pours no less than 1,000,000 metric tons of matter into the solar wind. This sounds like an enormous quantity, yet for our sun it represents no more than a harmless sort of "bloodletting." Throughout its existence, the solar wind has claimed less than 1/10,000 of the sun's total mass.

The expression solar "wind" is highly appropriate. It makes clear that we are dealing not with ordinary radiation, but with the emission of actual physical particles. Tiny though these particles may be, they form a wind which literally "blows" from the sun. Despite the incredible speed at which it travels, this wind is so thin and fine that it could not make a flag wave on the earth. Yet it is able to move something as fragile and ethereal as itself, such as a comet's tail. No doubt remains that the wind which makes comet tails turn like weather vanes is identical to the wind of protons and electrons that is blowing off the sun.

The Invisible Sphere

THE solar wind is actually the sun's own atmosphere expanding into space in all directions. It is no easy task to observe or photograph the sun's atmosphere. The only visible areas of the atmosphere are those which lie closest to the sun itself. Moreover, astronomers can photograph these areas only during solar eclipses or with the aid of highly specialized instruments. The areas near the sun form the fiery corona we have all seen in photographs (Illustration 18). The corona burns at a temperature of 1,000,000°C. It is much hotter than the sun's surface from which it emanates. Astronomers have not yet determined what causes this enormous temperature difference.

Astrophysicists have evolved one theory to explain the divergence in temperature between the sun's surface and its corona. The theory may sound bizarre, yet it accords with the laws of physics. Astrophysicists suggest that the temperature rise in the corona results from the bursting of so-called "granules" on the sun's surface. These granules are gigantic gas bubbles with an average diameter of some 600 miles, which are located deep inside the sun. The bubbles contain areas of turbulence moving at speeds of more than sixty miles per second. This inner turbulence causes the bubbles to rise from the sun's interior to its surface (see Illustration 19). The immense granules, whose temperature approaches 6,000°C, are continually bursting on the sun's surface.

These explosions must create an inconceivable din. Astrophysicists have used mathematical calculations to measure the level of the sound, and many of them believe that the sound waves released by the exploding bubbles heat the sun's atmosphere, thereby creating the energy to hurl the solar wind particles into space. Thus the sun seems to "fulfill its appointed round with the noise of thunder," as Goethe said—even if there will never be an ear to hear it.

In photographs, the corona may appear quite stable. Actually it is no more stable than a candle flame. Both corona and candle flame maintain their basic shape; but the matter composing them must be continually renewed. Despite the enormous size of the corona, it consumes its own substance once every twenty-four hours. Meanwhile the "candle," the sun, is feeding it new substance.

We have already noted that the particles of the solar wind are hurled into space with great force. Matter traveling at such high speeds must clearly travel long distances through space. When scientists discovered that particles of the solar wind were being hurled away from the sun, they realized that the sun's atmosphere did not really end with the corona itself. A candle flame also releases its substance into the space around it; meanwhile this substance is replenished from below, along the wick. Astronomers now faced the question of how far the corona extended into space. In other words, how far did the solar wind blow?

Scientists had observed the corona during the rare solar eclipses. Their instruments revealed fine traces of the corona at a distance of 10,000,000 miles from the sun. This may sound like a "far piece." On the other hand, at this rate the solar wind would not even have reached to the innermost planet of our solar system; Mercury is 36,000,000 miles from the sun.

Shortly after the last war, the development of radio astronomy created new investigative techniques. The huge parabolic antennae of radio telescopes are designed to locate and investigate so-called "radio sources" in space. These sources include fixed stars, nebulae, and galactic systems. All these bodies emit radio waves as well as visible light. Radio waves are much longer than waves of visible light, and have a low frequency; they lie far beneath the threshold of visibility. This new branch of astronomy has indirectly supplied scientists with data about objects which do not themselves emit radio waves. For example, radio astronomy supplies data regarding the sun's corona. Each year as the earth moves around the sun, the sun's corona interferes with our radio reception. Fine and thin as the corona is, it can scatter and break radio waves from various known radio sources (i.e., stars or galaxies which have already been tested with radioastronomical methods).

Scientists know the exact location of these radio sources. They can now measure how much the corona interferes with normal reception. As the earth travels around the sun, it continually moves to new positions in relation to the sun and to the various known radio sources. Thus the degree of interference with radio reception from each source varies according to the time of year. The new techniques of radio astronomy have enabled scientists to determine how far the corona hurls its disruptive particles into space.

In the 1950s astronomers used these sophisticated methods to "measure" the solar wind. The corona—or rather the particles it hurled into space—extended up to 44,000,000 miles from the sun. The orbit of Mercury lay well within this range. But beyond this point the solar wind became so thin that it could no longer be detected with radioastronomical techniques. Later the American Mariner space probes which traveled to Venus and Mars revealed that "solar plasma" (the more technical designation for "solar wind") actually extends far beyond the orbit of the earth and even beyond the orbit of Mars.

Thus not only Mercury, but also Venus, the earth, and Mars are fanned by the solar wind. Actually this "wind" is just the outermost layer of the sun's atmosphere expanding at great speed. What an astounding thought! In all probability our entire solar system lies *inside* the sun's atmosphere.

Thus the sun not only warms and illuminates the planets, it also envelops them in its own atmosphere. We might feel tempted to picture the sun as a mother hen spreading her wings protectively over her chicks. But the moment such a thought entered our heads, we would probably dismiss it again; after all, the image seems a little far-fetched and anthropomorphic. Yet in fact everything astronomers have learned in the last decade indicates that this far-fetched analogy is really quite appropriate! Without the protection afforded us by the solar atmosphere, our earth would be uninhabitable.

Why would we not survive without the solar wind? To begin with, we must return to our former question: How far does the solar wind blow? We have just noted that this wind extends *at least* beyond the orbit of Mars. Scientists know this from data relayed directly from the Mariner probes. Our space probes have not journeyed beyond the region of Mars, but there is no reason to assume that the stream of solar plasma stops flowing just beyond this point. When the solar plasma particles travel past the earth, they are still covering more than 186 miles per second. At this speed, the particles must hurtle on far beyond the orbit of Mars.

Scientists believe that they do. In fact, astrophysicists have been able to compute the probable extent of the solar wind in those regions not yet penetrated by our space probes. The Mariner probes recorded the speed and density of the "particle flow" in the regions of space near the earth. Astrophysicists based their calculations on these data. They also took into account those forces in space which might offer some resistance to the particles, slowing down the particle flow. Two factors are known to have a braking effect on the solar plasma. The more powerful of the two is interstellar matter. This matter consists primarily of hydrogen gas; but it also includes the finely distributed cosmic dust which occurs in irregular streaks throughout space. We know that this dust exists because we can photograph it; we can also demonstrate its effect on the light emanating from stars. In order to observe it, we must look several hundred light-years into space. Only at this distance does the "optical density" of the dust increase to the point that it becomes visible to our instruments.

Scientists are now able to determine the average density of the cosmic dust in a given region of space. They observe the appearance of a star passing through or behind the cosmic dust. The layer of dust dims the light of the star. The degree of alteration in its light tells scientists the density of the dust layer. For example, the dust layer might make the star's light look rather red. But how do we know that this reddish light does not simply represent the star's original color? How can we tell that the light has been "changed" by the dust? Actually, these questions pose no real difficulties. Spectroscopic analysis enables astronomers to determine a star's temperature and its true color. Once they have established the degree of color distortion, they can easily determine the density of the cloud layer which caused the reddening effect. To do so, they need only know the distance separating the earth from the star in question; then they can calculate the average concentration of the dust which fills the intervening space.

Interstellar matter is never very concentrated. It is composed of only about one atomic particle per cubic centimeter of space. This represents a thinner concentration than that of any artificial vacuum we can create on earth. Despite its almost incorporeal nature, this cosmic matter eventually acts as a barrier in the path of the expanding solar plasma. The plasma grows thinner in proportion to its distance from the sun; as it moves away from its source, it must expand farther in all directions. Space probes have measured the density of solar plasma as it streaks past the earth. Its density amounts to some five to ten particles per cubic centimeter. Using these figures, mathematicians can easily calculate how far from the sun the plasma will be when its density begins to equal that of interstellar matter.

Near the sun the solar wind is strong enough (several times the density of interstellar matter, with individual particles traveling at very high velocity) so that if any interstellar matter blocks its path, the solar wind will push it out of the way and keep going. In this way, the solar wind keeps the solar system swept relatively clear of interstellar matter. As the wind travels outward, however, it must spread out and decrease in density, until finally it reaches the point where it no longer has enough force to do this. At this point the solar wind is slowed down; it mingles with and becomes a part of the interstellar matter. One theory suggests that eventually the solar wind actually collides with the interstellar matter, creating a shock zone. In an earlier chapter we mentioned that two media whose atoms are tightly bound together (like my fist and a table) will not merge but will collide violently. The two partners in the collision can generate various forms of heat and turbulence at their point of impact. As they investigated the solar wind, some scientists surmised that if they discovered such areas of heat and turbulence in space, they might have found the point where the solar wind was halted in its course by its collision with interstellar matter.

Unfortunately the situation was far from clear-cut. Thus far we have not mentioned the *second* factor which exerts a "braking" effect on the solar wind. This factor somewhat complicated the astronomers' calculations. For they also had to consider the effects of the magnetic fields within our solar system.

For some time scientists had theorized that our solar system must contain weak magnetic fields similar to those that existed elsewhere in the universe. Space probes directly confirmed this theory. The magnetic fields of our solar system inhibit the progress of the solar plasma. The solar wind consists primarily of protons and electrons—in other words, of the fragments of hydrogen atoms. Everyone knows that in a whole atom the electrical charges of the protons and electrons are in balance. But the sun's heat "ionizes" atoms, that is, it breaks down the atoms into separate particles with individual electrical charges. The individual electrical particles of the solar wind are subject to magnetic influences.

It is difficult to judge how much interplanetary magnetic fields affect the solar wind. Scientists still know relatively little about the strength of these fields, their location in space, and their patterns of movement. Nevertheless, astronomers have succeeded in setting up some general guidelines. They have established an upper and a lower limit for the possible extent of the solar wind. They achieved this by setting two hypothetical cases. First, they computed the *maximum* degree to which inhibitory factors might slow down the solar wind. Second, they calculated the *minimum* possible influence of these factors. By this method they discovered two things: the smallest possible distance that the sun's atmosphere *must* extend into space, and the greatest possible distance that it *can* extend.

Astronomers repeated their calculations many times over. These calculations indicated that the sun's atmosphere must extend at least one billion miles into space, which would carry it far beyond the orbit of Saturn. On the other hand, it *might* have a radius of no less than 15,500,000,000 miles; in this case its diameter would equal more than *four times* that of our solar system.

Probably neither estimate hit the nail exactly on the head.

More than likely the truth lies somewhere between the two extremes. Astronomers now suspect that the solar atmosphere envelops the entire solar system. For billions of years the sun's gravitational attraction has bound our nine planets together; it has warmed and illuminated them all. But beyond all this, the sun has sheltered the planets inside its own atmosphere—an atmosphere composed of the solar wind that streams outward beyond the orbit of Pluto.

As we have shown, space probes are helping us form a new picture of our solar system. The sun's atmosphere fills all the space of this solar system. This fact means that interplanetary space is not empty. Moreover, interplanetary space is divided into distinct "zones."

Strictly speaking, our probes have not yet penetrated "outer space." Even the unmanned flights to Venus and Mars did not reach what could truly be called outer space. Apparently our whole solar system exists in a special milieu which ends somewhere beyond the orbit of Pluto. The sun's atmosphere creates this special milieu. Beyond Pluto, a separate zone begins—the zone of outer space.

If it were not for the special milieu created by the sun's atmosphere, there would be no life on earth. To understand this fact, we must first form a picture of the border zone separating the sun's atmosphere from the outer space beyond. We have already noted that the solar atmosphere may end at a kind of "shock zone," where the thinning solar plasma comes into conflict with interstellar dust. Such a shock or border zone would form a huge sphere surrounding our entire solar system. The magnetic storms created by the clash of solar plasma with interstellar dust would arise on all sides at an approximately equal distance from the sun.

If we recall our earlier measurements, then it will be clear that this sphere might have a diameter in the neighborhood of some 7 to 9 billion miles. The "walls" of the sphere, formed by the turbulent areas at the border zone, would be somewhere between a few hundred and a few thousand miles thick. Thus our entire solar system may be enclosed within a huge sphere. This sphere would not only be invisible, but also incorporeal. The solar particles are not nonmaterial, but the magnetic storms they may create *are*. We may all be living under the protection of an invisible and nonmaterial force.

The earth is constantly bombarded by so-called "cosmic radiation" from outer space. This radiation was discovered by accident around the turn of the century and is still giving physicists quite a few headaches. Cosmic radiation represents the most intense form of radiation ever measured. Moreover, no more intense form of radiation could possibly exist. The particles composing it move at almost the speed of light. (Nothing in nature can exceed the speed of light.) In addition, cosmic radiation can penetrate solid matter, passing through lead walls several yards thick. Nor is it deterred by hundreds of yards of solid rock; cosmic rays have been recorded even in very deep mines. Fortunately this intense radiation is very finely diffused; swift as they may be, its particles are few in number. If this were not the case, the earth's surface would no doubt be a pretty uncomfortable place.

For decades scientists had known about this cosmic radiation. Then the investigation of the solar wind cast new light on the subject. It turned out that cosmic rays were not quite as harmless as previous measurements had indicated. The measurements had been made from the earth. Instruments on earth had not recorded the fact that the outer space beyond our solar system must be filled with the raging of cosmic radiation, only a few particles of which ever reach this planet. Such particles have succeeded in crossing the thin barrier around our solar system created by the solar wind. This invisible sphere is all that lies between us and the deadly radiation that bombards us from all sides at nearly the speed of light.

Even assuming that the solar system is protected by a magnetic sphere in addition to the solar wind, this sphere seems so fragile that we may wonder how it can ward off the most intense radiation that exists. But the seeming contradiction resolves itself as soon as we see how the protective sphere would really function. The spherical shock zone would not actually halt the progress of the radiation: after all, this thin vacuum-like border zone could not achieve what defeats even lead walls several yards thick. Instead, invisible, nonmaterial magnetic storms created by the turbulence of the electrical particles would apparently act as a sort of "reflector." The radiation particles are far too powerful to be intercepted or slowed down. Instead they may meet with a kind of elastic resistance; in this case lines of magnetic force would scatter them and divert them onto a new course. The giant sphere beyond Pluto would not act as a solid wall. Instead it would behave like a mirror which gathers the forces from outer space and sends them gliding outward again.

We could describe the new picture of our solar system in this way: The sun "blows" so powerfully in all directions that it creates a spherical space impervious to cosmic rays. This spherical space is so vast that the entire solar system fits inside it. Possibly the solar wind collides with interstellar matter, thus creating a turbulent border zone. The turbulence in turn would produce magnetic storms. These facts were deduced from data relayed by our space probes; the data revealed to scientists the basic properties of solar plasma in the region of the inner planets. We also know that the magnetic fields of the border zone, if in fact they exist, would hinder cosmic radiation from entering our solar system.

We now have direct evidence of the protective properties of the spherical border zone. This evidence consists in the so-called "Forbush effect." Scientists have long been aware of this phenomenon, but they did not understand it until very recently. New discoveries regarding the probable existence of a magnetic border zone have clarified the nature of the Forbush effect. This puzzling phenomenon was named after the scientist who discovered it. At times a sudden decrease occurred in the amounts of cosmic radiation recorded on earth. Several days passed before the radiation manifested itself again with its usual intensity. Although the reductions in the radiation level occurred at irregular intervals, scientists had noticed that these reductions seemed related to periods of eruption on the sun. More precisely, a marked decrease in radiation always occurred a few days after a solar eruption. The two phenomena were somehow related; but until recently no one had any idea what the connection might be. Now the problem seems solved.

The Forbush effect fits perfectly into our newly acquired picture of the dynamic relations between our solar system and the outer space beyond it. In the context of this picture, we can easily understand why local eruptions on the sun's surface should cause a temporary reduction in the levels of cosmic radiation on earth. A solar eruption increases the number and speed of the particles being hurled out of the corona, thereby temporarily increasing the power of the solar wind. In turn, the magnetic border zone must become a more effective shield. In other words, the Forbush effect clearly reveals the protective role of the spherical border zone.

Now it is clear why our sun deserves to be called the "life-giving star." The sun provides Spaceship Earth with light and energy. Equally essential to our lives is the energy the sun so extravagantly pours into "empty" space. The solar wind is as necessary to our existence as the tiny fraction of the sun's light which actually reaches the earth.

If the sun were suddenly extinguished, we would not have time to freeze to death—we would die from the effects of cosmic radiation first. Huge reservoirs of heat are stored in our atmosphere and in the earth's crust. Long before these were used up, we would have succumbed to the deadly rays pouring down on an unprotected earth.

But our story is really only beginning. The new discoveries regarding our solar system and outer space have raised a number of questions, many of which scientists have not yet had time to answer. But one thing is clear: a striking picture is emerging, forcing us to revise our previous views of the earth's role in the universe.

A couple of years ago, who would have dreamed that we were living inside a giant invisible magnetic bubble which extends far beyond the orbit of Pluto? If we pick up and follow this thread, we will find many more surprises lying in store for us. Our earth forms part of an incredibly complex network of cosmic interrelationships. For example, we shall soon see that the earth would be uninhabitable without the moon. We may seem to proceed rather slowly, but we must take care not to overlook the essential and fascinating details of the earth's journey through space.

A few years ago an experiment was conducted in the Sahara Desert. Since then, the same experiment has been repeated in various other regions of the earth. Compared to the satellite launches, this experiment attracted very little public attention. This was hardly surprising, since only scientists would find it particularly significant. An observation rocket was launched into the upper levels of the earth's atmosphere. At a height of around 125 miles, the rocket released a small cloud of barium. This cloud remained under telescopic observation; it was repeatedly photographed in both color and black and white. If we examine these photographs in their proper sequence, we see that the cloud was behaving very strangely. It was beautiful to look at, since it was floating in the so-called "ionosphere"; the sun's rays made it glow with a sort of fluorescent light. Then the cloud began to expand uniformly in all directions, so that it slowly grew larger without losing its original spherical shape. Up to this point it was behaving exactly as predicted.

But several minutes later the cloud seemed to divide into two separate substances, each of which was behaving quite differently from the other. Part of the barium continued to expand uniformly in a roughly spherical shape. At the same time a second cloud began to separate from the first. This second cloud was slightly different in color and had a cylindrical shape—it was expanding at both ends of its north-south axis. How did scientists explain this bizarre phenomenon?

The explanation proved quite simple. At this altitude the atmosphere contains only around 5 billion atoms per cubic centimeter. (At sea level the proportion is 2.5 times 10^{19} per cubic centimeter, which gives us a number with eighteen zeros—twenty-five quintillion.) The sunlight easily penetrates an atmosphere with such a small atom population. The sunlight not only causes the barium to glow, but also partially ionizes it. A barium atom which has undergone ionization loses one of its electrons. Thus two different substances with different properties develop inside the cloud. The nonionized portion of the cloud continues its spherical expansion, obeying a simple mechanical law. The ionized portion, on the other hand, glows with a somewhat different color. Moreover, being composed of particles whose electrical charges are no longer in balance, it becomes subject to *magnetic* influences. Its cylindrical expansion toward the north and south reveals the presence of a force which normally remains invisible, the earth's magnetic field. The barium cloud experiment in the Sahara was designed to observe and measure the behavior of this magnetic field at various atmospheric heights.

Scientists had several reasons for wishing to learn more about the geomagnetic field. For one thing, they were looking for the answer to a question relating to the solar wind. The question was this: We know now how we are protected from radiation by the solar wind, but what protects us from the solar wind itself?

It is fascinating to visualize how the sun shelters its dependent planet family in its own atmosphere, shielding it from cosmic radiation. But the protection it affords consists of electrical particles traveling at speeds of several hundred miles per second. It may thus appear that we have more or less been caught between the devil and the deep blue sea. After all, are we not in just as much danger from the solar wind as we could be from cosmic radiation? To be sure, by the time the protons and electrons of the solar wind have reached the earth, they are traveling at only about 186 miles per second. Therefore they cannot be nearly as destructive as the cosmic radiation, which travels at almost the speed of light. All the same, we are constantly bombarded with particles moving at a thousand times the speed of sound. These particles strike us from a distance of only 93,000,000 miles away. The continual bombardment must have some negative effects. (In a later chapter we will encounter some examples of these negative effects.)

The "corpuscles" of solar plasma bleeding outward from the sun shield us from cosmic radiation. But what protects us from the solar plasma? It may seem astonishing that we are protected by a frail force barely strong enough to pull a compass needle toward the north—the earth's magnetic field!

A Cage for the Solar Wind

MAGNETIC force is invisible. Thus we cannot see the pattern formed by the earth's magnetic field. Nor can scientists trace the field in its entirety; they must measure it in individual segments. Lines of magnetic force can be measured at individual points with the aid of a compass or other appropriate instrument, or through experiments like that in the Sahara. Magnetic force fields form a dense network all around the earth. If we could actually see magnetic fields, the earth would look as if it were enclosed in a giant cage made of lines of magnetic force. This cage would consist of two half spheres of equal size. The earth behaves much like a bar magnet whose northern and southern poles more or less coincide with the earth's axis of rotation. The lines of force composing the magnetic field originate at the two poles. This fact enables scientists to trace the general pattern of the field. The lines of force rise almost perpendicularly from the poles, gradually curve along a horizontal line, and finally descend in an almost perpendicular line to the opposite pole (see diagram).

We must now ask a question which may at first seem rather academic: How far does the earth's magnetic field extend into our atmosphere? Recent space probes have given us an intriguing answer to this question. The same space probes that disclosed the nature of the solar wind revealed that the solar wind and the geomagnetic field each have their separate realm in space. Physicists investigating both phenomena were not really surprised to discover this territorial division. After all, the solar wind consists of electrically charged particles susceptible to magnetic influences. The earth's magnetic field "controls" the solar wind by simply shutting it out. Satellite investigations showed that the magnetic field contains almost no particles of solar plasma. (There is one exception to this rule, which we shall discuss later.) Eventually the magnetic zone free of solar wind came to be designated as the "magnetosphere."

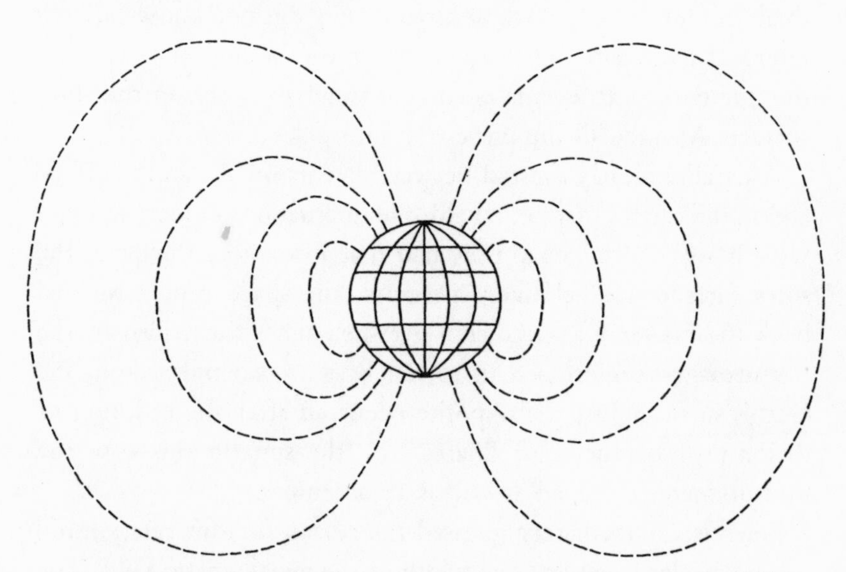

Pattern formed by the lines of force in the earth's magnetic field.

The magnetosphere is the area above the earth's atmosphere controlled by the magnetic field and impervious to the solar wind. In other words, the magnetosphere is the plasma-free area surrounding the earth. We have already noted that the earth behaves very much like a bar magnet. Therefore scientists at first assumed that the magnetosphere, like the atmosphere, would turn out to be a genuine "sphere," that it would literally have the shape of a ball. But further satellite investigation revealed a quite different and far more dramatic picture.

Using space probes, scientists tried to determine the true dimensions of the magnetosphere. They wanted to know how far its outer edge lay from the earth. But the probes kept recording contradictory figures. Sometimes the data they transmitted showed fluctuations of almost one hundred percent. At first scientists suspected that these data might simply represent errors in measurement. But eventually it became clear that the upper and lower limits of the space probe measurements remained constant. True, the individual figures continued to fluctuate, but these figures were never lower than around 27,000 miles, never higher than around 50,000. At first astronomers did not know how to interpret these data. Then someone hit on the notion of relating the fluctuations to events occurring simultaneously on the sun's surface. At once all the pieces fell into place.

A parallel clearly existed between events on the sun's surface and in the earth's magnetic field; the fluctuations in measurement were not as capricious as they had first appeared. Whenever the sun's surface was relatively peaceful, the space probes relayed back the *higher* measurements. At such times the width of the magnetosphere increased to as much as 50,000 miles from the earth's surface. Just the opposite occurred after the eruption of major protuberances, or "flares," on the sun. In this case the measurements dropped as low as 27,000 miles.

Scientists immediately guessed the reason for this relationship between solar flares and the width of the geomagnetic field. The invisible magnetic cage surrounding the earth shields us from the solar plasma; as it does so, it trembles and quakes under the heavy bombardment of solar wind particles like a soap bubble caught in a draft. Periodically eruptions occur on the sun's surface, the so-called "flares" (see Illustration 17). These flares temporarily intensify the particle flow of the solar wind. The solar wind in turn deforms the spherical shape of the magnetosphere, forcing its outer rim some 15,000 to 20,000 miles back toward the earth. In other words, the sunward side of the magnetosphere "flickers" under the fluctuating pressure from the solar wind. The "flickering" variation in the width of the magnetosphere may equal two to three times the total diameter of the earth.

We must keep in mind that all these processes are absolutely invisible. Only the plasma corpuscles borne by the solar wind are material in nature. A person looking out the window of a spaceship would see nothing but empty space. Here we have an especially striking example of a truth that we human beings rarely consider: Our sensory organs perceive only a fragment of reality. We are equipped to grasp only that dimension of nature which we *must* grasp in order to fulfill our biological role. We all too readily forget that our perceptual range is incommensurate with the total reality of the world. Thus the entire magnetosphere and its confrontation with the solar wind bombardment remain invisible to us. This drama becomes accessible to human beings only through the artificial sensory organs built into our modern space probes. And yet nothing could be more "down to earth" than the processes daily taking place in space. Every day the magnetosphere protects us from the solar wind just as the solar wind shields us from cosmic radiation.

Thus we live under the protection of two invisible spheres, one fitting inside the other. The first sphere is generated by the solar wind; it consists of the shock zone extending beyond the orbit of Pluto, that owes its existence to the clash between solar plasma and interstellar matter. It envelops our entire solar system. The second sphere is much smaller but equally important to human beings. It consists of the magnetosphere, which shields the earth from solar plasma.

Let us turn once more to this second sphere and to the lively activity taking place on its surface. We have already noted the 15,000- to 20,000-mile fluctuations in its sunward half. Recent space probes have revealed other remarkable things about the magnetosphere.

The fluctuations in the "front," or sunward, side of the earth's magnetic field are clearly produced by pressure exerted by the solar wind. When scientists first explained this phenomenon, they found supporting evidence in the fact that the fluctuations grew less pronounced toward the *sides* of the magnetosphere. These lateral areas were less exposed to the solar wind, and remained almost constant in width. Regardless of conditions on the sun, their width was always recorded as about 55,000 miles. From this lateral angle the solar wind only grazed the magnetic border zone. Therefore the spherical shape of the magnetosphere was less distorted here than at other points. In other words, the flanks showed most clearly what shape the magnetosphere would have had if the solar wind had not existed.

We have already noted that the sunward side of the magnetic field never exceeds a width of about 50,000 miles. The flank areas, on the other hand, maintain a constant width of some 55,000 miles. This discrepancy in width shows that the solar wind blows constantly. Even when the sun's surface remains relatively peaceful, the solar wind blows hard enough to push the magnetosphere more than 5,000 miles back toward the earth.

One might assume that, like the flank areas, the rear of the magnetosphere would be quite stable. After all, the rear area is shielded from the sun by the intervening bulk of the earth. That is, this area appears to be completely out of reach of the solar wind. Yet actually just the opposite is true; the deformation of the magnetosphere turns out to be *most* acute on its "sunless" side. Moreover, conditions here are far more turbulent than elsewhere in the magnetosphere. The turbulent conditions greatly complicated scientific investigation of the region. In time astronomers discovered the reason for the turbulence. They learned that the solar wind behaves very much like the winds we know on earth. The side of the magnetosphere turned away from the sun is "picked up" by the solar wind and blown out into space. Its frayed tatters may be carried more than 600,000 miles from the earth. Thus the magnetosphere literally flutters in the solar wind as a candle flickers in a draft.

This fluttering or flickering effect was first detected by the Mariner IV probe toward the end of 1964. During the first part of its historic flight to Mars, Mariner passed through the rear of the magnetosphere, which is turned away from the sun. Mariner continued straight on its path. But within an hour's time the probe passed through the same magnetosphere no less than six times in a row! Mariner was traveling at about 2,500 miles per hour. But the tail of the magnetosphere kept fluttering in the solar wind, and repeatedly caught up with the retreating probe. Let us try to picture how this whole process looks. Only then can we gain some idea of the true power of those silent and invisible forces raging all around us.

If we could actually *see* magnetic fields, the earth would not really resemble the earth in our earlier diagram. In other words, the earth would not really look as if it were enclosed in a cage formed of two half spheres. Its actual appearance corresponds to this second diagram, which shows the earth's exposure to the solar wind. The earth drags a tail behind it very much as a comet does. The earth's tail consists of lines of magnetic force. Moreover, the same force creates both the earth's tail and the comet's tail—the pressure of the solar wind. This pressure also determines the direction in which the tail will stream out in space.

The earth's magnetic tail also proves interesting because it casts some light on the Van Allen radiation belts. Anyone who has followed our explanations up to this point will have been struck by an apparent contradiction, one which must be cleared up before we can conclude our discussion of the relationship between

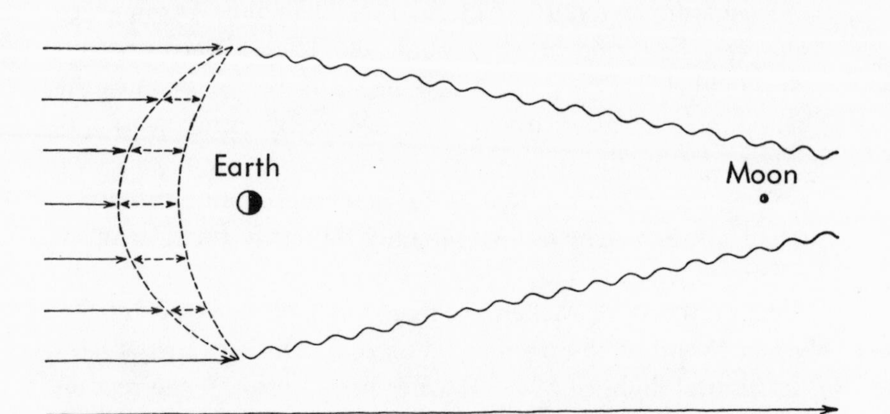

The earth's exposure to the solar wind. On this scale, the sun would be a sphere about 18 inches in diameter, located nearly 55 yards to the left of the diagram. The protons and electrons of the solar wind originate in this sun and reach the earth at a speed of 186 miles per second. They are halted by the magnetosphere. The impact forces the sunward side of the magnetosphere to retreat some 15,000 to 20,000 miles back toward the earth (two sets of dotted lines). The side of the magnetosphere turned away from the sun behaves like a flickering candle in the path of the solar wind; it is drawn out into a long tail which extends past the orbit of the moon (wavy lines).

the magnetosphere and the solar wind. We have defined the magnetosphere as a plasma-free area surrounding the earth, and we said that the magnetic cage shields us from the solar plasma. Yet we have also spoken of the two Van Allen radiation belts which lie deep within the magnetosphere. Both belts are composed of protons and electrons which originally traveled to the earth in the form of solar plasma. How could these particles have slipped through our magnetic defense system?

We have not yet found a truly satisfactory answer to the question. Probably the breakthrough of solar plasma takes place through the rear, or "tail," of the magnetosphere. So much turbulence exists in this area that it no longer maintains a solid barrier against the solar wind. Thus a few particles literally infiltrate our defenses from the rear.

But even these few particles fail to reach the earth's surface. They are halted by a second line of defense composed of additional magnetic fields. These rearguard magnetic forces are more compressed, and therefore stronger, near the earth; they occupy two zones located around 4,000 and 12,000 miles from the earth. The magnetic zones capture the intruders and imprison them in what physicists call a "magnetic bottle." These two zones are none other than our old friends, the Van Allen radiation belts! They both act as reservoirs for storing solar wind particles which have penetrated our defenses.

Yet these trapped particles remain active and therefore dangerous inside their magnetic prison. The radiation belts do not reduce the energy of the particles. The protons and electrons continue to race about at almost undiminished speeds, although now they move along corkscrew paths between the earth's northern and southern poles. They are moving so fast that they need hardly more than one second to make a round trip from the North to the South Pole and back! A single particle may sometimes spend weeks or months in one of the radiation belts before escaping into the outer atmosphere through a gap in the polar regions. The lower belt is more stable than the upper; a particle may survive here for years before it gets away! The escape of particles through the polar gaps occurs during temporary reductions in the width of the magnetosphere, or during especially powerful solar eruptions. These "escapes" are marked by interference with radio transmission on earth and by brilliant displays of the polar lights.

Thus Olaf Birkeland's almost eighty-year-old theory has been proved amazingly accurate. Yet Birkeland and his contemporaries did not suspect how complex the situation really was.

We have now discussed a number of cosmic relationships. We began by noting that the earth is not an autarky, independent of everything outside it. Its life cycles are maintained by energy supplied by the sun. The quantities of energy required by the earth over vast periods of time can be supplied only by thermonuclear reactions.

We have further seen that the sun's vast size fulfills two important functions for our earth. First, this enormous bulk builds up the temperatures and pressures needed to set off atomic reactions in the sun's core. Second, this bulk acts as a giant filter; it reduces the intensity of the sun's internal energy so that this energy does not harm living creatures on the earth.

A few observations seem appropriate at this point. We should not make the mistake of thinking that the sun developed these beneficial properties "in order that" life could develop on earth. Such a notion would presuppose that some external power had purposely arranged everything for our benefit. Actually just the opposite is true. The sun developed long before the earth and earthly life even existed. It certainly never tried to adjust itself to our future circumstances. Instead, life on earth adapted to the given conditions, which were largely determined by the properties of the sun. This fact does not make the development of life any the less astonishing. All the same, some people tend to view the whole situation the wrong way around, betraying a narrow and anthropomorphic point of view.

Despite the sun's beneficial effects, it contains particles which could endanger life on earth, the energy particles of the solar wind. Strangely enough, these same particles also play a part in the sun's life-giving role. The solar wind gives rise to that sphere which shields us from deadly cosmic radiation. Our magnetic field forms a second sphere which protects us from the solar wind.

Thus far we have seen that our earth is not isolated in space, but participates in a complex and beautifully crafted network of cosmic interrelationships. Nor is our earth the stable phenomenon our everyday surroundings lead us to believe. Our entire solar system consists of a delicate balance of powers; one might compare it to a giant cosmic mobile suspended in space.

Before we come to the end of our tale, we must examine many other fascinating scientific discoveries. Some of these concern our moon, which belongs to the network of cosmic relationships that maintains our life. We shall find out more about the moon in the course of answering our next question: What is the origin of the earth's magnetic field?

A Planet Made Transparent

At some time during our lives we have all looked into a pond or lake on a windless day when the surface of the water was as smooth as a mirror. We noticed something that everyone notices sooner or later—in fact, it is such a common experience that we all take it for granted. Yet most people would be hard put to explain it. Directly in front of us we could see right through the surface of the water to the bottom. We saw fish swimming around. But a little farther away the water again became an impenetrable mirror; distant objects like clouds or the opposite bank were reflected in its surface.

Waves (in this case waves of light) are refracted when they move from one medium to another—they are deflected from their original course. They may pass from a thin medium into a denser one, or vice versa. In this case the light waves transfer from water to air. The angle of refraction depends on various factors, such as the wavelength of the refracted ray and the properties of the two media through which it travels. But above all it depends on the angle at which the ray strikes the surface between the two media (in this case the surface of the water). The wider this angle, the greater the degree of refraction. Eventually a critical point is reached at which the light ray is no longer refracted, but is reflected instead. The ray behaves like a bullet which strikes a hard surface at too wide an angle: It does not penetrate the surface but ricochets from it and flies back into the medium from which it came.

All this belongs to elementary physics. We mention it here simply because it clarifies the principle used by geophysicists in investigating the inner structure of the earth. Many regions have seemed off limits to human beings—the polar regions, the dark ocean depths, or extraterrestrial space. With the aid of technology, we have discovered how to gain access to all these realms. But one region very close to us will probably remain sealed away from human beings forever, the earth's interior. True, we do sometimes drill a short distance into the earth's crust. One day perhaps we may even penetrate to the deepest regions of the mantle. But pressure and temperature markedly increase below the mantle; it is futile to dream that men will ever descend beneath this level.

Despite our inability to directly penetrate the earth's interior, geophysicists have discovered a way to uncover its secrets. The earth's crust averages some twenty miles in depth. The next layer, the mantle, extends nearly 1,900 miles down. Then comes the outer core, composed of some 1,300 miles of liquid nickel and iron. The final layer is a hard core of solid nickel and iron, around 1,500 miles in diameter.

Scientists can come within one percent of determining the exact dimensions of structures inside the earth. They have discovered a way to literally "see inside" these impenetrable depths. But they do not "see" with light. We have noted that a surface becomes visible when it "refracts" waves, that is, when it deflects them from their original course. But it also becomes visible by *reflecting* the waves. This is the experience we have at the edge of a lake. When we are looking almost perpendicularly down at the water, its surface remains invisible; we see the lake bottom and the fish, but we look straight through the water. Just the opposite happens when we look toward the center of the lake. Our visual angle is now so wide that the bodies of the fish become distorted by the refraction. Thus we indirectly grow aware of the surface of the lake. Eventually our angle of vision has grown so wide that the light rays no longer penetrate the water, which now appears as a reflecting surface.

All this applies to any visual experience we may have. We can only see an object which retards or at least distorts light rays, as waves of hot summer air are distorted above an asphalt pavement. Sometimes people in fantasy novels have the power to make themselves invisible. They go through any number of bizarre or amusing experiences. But the authors of these fantasy tales always overlook something that should have been quite apparent to all their invisible characters: Anyone who was truly invisible would also be totally blind! Such a creature of fantasy could achieve invisibility in only one of two ways. He could develop a technique to make light rays flow around his body; then only the objects directly behind him would appear undistorted. Or he might take a miraculous drug which would render his whole body completely transparent. In either case, this invisible man would be unable to see a thing. If he used the first method, not one single ray of light would reach his eyes; the light would all flow *around* him. In the second case, light rays would pass straight through his eyes. In order to see anything, the eye must receive and refract light rays and convert them into nerve impulses. If the invisible man wanted to see anything, he would have to allow at least his eyes to remain visible. These disconnected eyes would make for rather a grotesque sight, and would no doubt defeat the whole purpose of the enterprise. After all, invisible people are usually trying to escape notice, not attract it!

All the same, creatures actually exist which are virtually transparent except for their eyes. (They should convince the skeptic that this discussion is not quite as fanciful as it may sound.) These transparent creatures are forms of deep-sea life, particularly jellyfish. Their glassy bodies are invisible in water except for their light-sensitive eyes, which are always solid in appearance. These eyes look like black or brightly colored beads hovering all alone in the water. To be visible, an object cannot be transparent, that is, it must be capable of refracting light rays. Anyone who has ever walked right into a glass door can testify to this fact.

The waves which geophysicists use to "see" inside the earth are seismic waves. These are mechanical waves, relatively simple in comparison with the electromagnetic light waves. All the same, they possess the same properties and obey the same laws as we have just described in connection with light. They resemble light rays in that they undergo refraction in passing between two media of unequal density. They also resemble light in their ability to create a reflecting surface by striking this surface at a wide angle.

Besides seismic waves, scientists know another method to achieve the same effect: They can also irradiate the globe using waves created by artificial explosions. Such waves offer the advantage that one knows exactly when they will appear. On the other hand, even underground atomic explosions create waves that are very weak in comparison with seismic waves. (For example, one severe earthquake produces the energy of at least 100,000 atomic bombs of the type dropped on Hiroshima.) Each year more than 100,000 earthquakes occur all over the earth. Fortunately most of them are too weak to cause serious damage. But they produce more than enough seismic waves for research purposes.

The speed at which these waves spread through the earth depends on the density of the material through which they are traveling. They move most slowly in water, around one mile per second. Their speed through granite averages three to four miles per second; through the hardest layers of the mantle, more than five miles per second. Seismic waves travel most swiftly through the earth's core—about seven miles per second. Some waves are strong enough to traverse the entire diameter of the globe, passing right through the center of the earth. And it takes them only twenty-two minutes!

Seismic waves enable scientists to determine the location of layers of various density in the earth's interior. The speeds at which the waves travel through the various layers reveal the physical properties of these layers and their probable chemical composition. Waves from each quake are clocked at various points on the earth's surface. The time intervals between the various checkpoints indicate the speed of the waves and their degree of refraction.

There is a second type of seismic wave which scientists discovered through laboratory experiments and theoretical calculations. This second type of wave can travel through solid matter but not through liquids. For example, an earthquake may occur in New Zealand and be recorded by a seismographic station in England. If the station records only the "primary" waves, this indicates that a liquid zone must exist in the earth's interior somewhere between England and New Zealand.

Investigations of this sort have proved that the outer layers of the earth's core must be liquid. These layers of liquid iron and nickel extend from around 1,800 to 3,200 miles beneath the earth's surface. Below these levels lies the true iron and nickel core of our globe. Here the metals are probably solid. For the purposes of our discussion, the most important fact scientists have established about the earth's interior is the fluid nature of its outer core.

We had reached the point of asking how the earth acquired its magnetic field. We can by no means take it for granted that such a magnetic field should exist. Recent space probes have informed us that our nearest neighbors in the solar system, Venus and Mars, have no magnetosphere. Thus the surfaces of these planets are totally unprotected from the solar wind.

Human beings have long been familiar with various magnetic effects. Ever since the Chinese invented the compass, sailors have been using it to find their way at sea. Then in the year 1600 the English physician William Gilbert published his *De Magnete*. Gilbert was the first to express the view that the whole earth might be just one giant magnet. Later astronomical and geophysical research gave us clues to the true nature of the earth's interior. These new data supported Gilbert's theory that the earth as a whole possessed magnetic properties. Scientists thought they had found a simple explanation for this phenomenon. By observing the moon's orbit, astronomers were able to compute the degree of gravitational attraction which the earth exerted on the moon. On the basis of this computation, they determined the earth's total mass. They related these figures to known data regarding the weight and composition of the earth's crust. In the end, scientists concluded that the earth's interior must contain great quantities of iron, since otherwise the earth's mass would have been disproportionate to its known volume. Everything seemed clear so long as one assumed that the metal inside the earth had magnetic properties.

Scientists were satisfied with this explanation for more than a hundred years. Actually it must have disconcerted them a little to find that the earth's axis of rotation almost exactly coincides with the axis of its magnetic field. If the earth had simply contained a huge iron magnet, then its magnetism and its rotation ought properly to have been unrelated. But the final refutation of this oversimplified theory came from another quarter. The theory was disproved by the discovery of the high temperatures in the earth's interior.

Drilling operations and the construction of mining shafts revealed that the temperature of the earth's crust increases rapidly with increasing depth. Near the surface the temperature rises only about 3°C for every 330 feet of depth. To be sure, this temperature rise must level off after a certain distance, since otherwise the resultant temperatures would contradict other known data. Seismic waves demonstrate that the matter composing the earth's mantle remains quite solid at depths of 1,800 miles. (Of course, local lava sources in active volcanoes form an exception to this rule.) But this solid matter could not be solid if the temperature continued to rise at the rate of 3°C per 330 feet. If this rate continued, the temperature in the mantle would be almost 100,000°C, and the mantle itself would be in a liquid state.

Our knowledge of temperatures inside the earth is not especially reliable. We base our temperature estimates on two factors: calculations of the pressure at various levels, and measurements of the quantities of heat produced by the deterioration of radioactive elements inside the earth. We now estimate the temperature at a depth of 25,000 miles to be 1,000°C; at 1,800 miles, 3,000° to 5,000°C; at the earth's center (about 3,800 miles beneath the surface), some 10,000° to 12,000°C. Inexact as these figures may be, one thing becomes clear: At a depth of only twelve miles, the temperature will already have reached about 775°C. Above this temperature, iron completely loses its magnetic properties. The "bar magnet" theory of the earth's magnetic properties (the theory that the earth's core was largely formed of metal with magnetic properties) eventually proved untenable. Continuing investigation has showed that the problem of geomagnetism is far more complex than was at first assumed.

In time additional evidence was found to refute the old magnetic theory. This new evidence proved that the earth's magnetic field was not created by any static or permanent source. If there had been some form of permanent magnet inside the earth, no fluctuations would have occurred in the strength of the magnetic field. In reality, its strength fluctuated to varying degrees at irregular intervals. In other words, the magnetic field was clearly created by some unknown *process* in the interior of the earth.

This was the point of origin for the now widely accepted "dynamo theory." The American physicist Walter Elsasser was largely responsible for its development. According to the known laws of physics, only one method exists of producing a nonpermanent magnetic field—the creation of magnetic force by means of electrical current. The question then was: Where inside the earth are there electrical currents capable of generating our magnetic field?

Electrical current can flow only in materials which act as good conductors. Metal makes an excellent conductor; but so does ionized gas. Scientists now believe that the motion of gas ions in our upper atmosphere (the so-called "ionosphere") contributes to the creation of the geomagnetic field. To be sure, it contributes only a few percent of the total magnetic force. Inside the earth itself, electrical current could be conducted only by metals—above all, by the huge quantities of nickel and iron which make up the core.

At least the outer shell of the earth's core is liquid. The earth's interior could create the magnetic field only if this liquid part of the core were to act as a huge generator for the production of electrical current. The flow of the current would then generate the magnetic field. The estimated temperature of the iron in this region is some 4,000°C; this hot iron moves at a rate of only one to two yards per hour!

From the physicist's point of view, the dynamo theory displays a high degree of internal consistency. This theory also explains why the earth's magnetic poles should coincide with its geographic poles. The lines of the magnetic field rise from the earth's surface at just the point where the earth's axis of rotation comes to an end (i.e., at the two poles). The dynamo theory casts some light on this hitherto puzzling phenomenon. We can assume that the flowing motions in the earth's liquid core originally formed isolated whirlpools or areas of turbulence. These turbulent areas created currents which in turn created various magnetic fields. There were many of these fields of differing sizes scattered at random throughout the earth. The unified magnetic cage to catch the solar wind did not exist; instead, the map of our magnetic field resembled a huge patchwork quilt. Yet we now perceive the magnetic field as unified and symmetrical in shape. Only the pressure of the solar wind is able to somewhat modify this shape. But what could have created this symmetrical form? The dynamo theory provides a ready solution: The present unity of the magnetic field is simply the result of the earth's rotation! The constant rotation of the globe smooths out all the whirlpools at the core. As the earth turns, the whirlpools begin to flow in a single stream. All the flowing motions in the core begin to move in one direction. They transmit a single current which rotates around the earth's axis.

The liquid and metallic character of the earth's core is one prerequisite for the development of the magnetosphere; the rotation of the earth is another. Great quantities of matter and energy go into keeping us alive and shielding us from solar plasma. But something else besides matter and energy is needed to produce a magnetic field. Let us once again consider the question of why Mars and Venus lack a magnetic field. If the theory we have just outlined really explains the existence of our magnetic field, then it should also explain why our two neighbors failed to develop one.

The case of Mars seems to substantiate the dynamo theory. Our neighboring planet is orbited by two moons, Phobos and Deimos. It is about half the size of our earth (the earth's diameter is about 7,900 miles, that of Mars about 4,200). But Mars has only about ten percent of the earth's mass and only seventy percent of the earth's average density. In other words, in all probability Mars lacks a metallic core like the earth's. Moreover, the low mass of Mars would probably not produce sufficient internal pressure to liquefy whatever metal deposits might exist in its core.

Thus the dynamo theory easily explains why Mars should have no magnetic field.

Venus is not quite so cooperative. Venus almost equals the earth in size (diameter 7,700 miles), with only twenty percent less mass. Thus it approaches the density and weight of the earth. Since roughly the same conditions prevail on Venus as on the earth, we must assume that Venus too has a partially liquid metallic core.

Yet Venus has no magnetic field. One possible explanation lies in this planet's remarkably slow rotation. Until recently we knew nothing about the rotation of Venus. Its atmosphere is so clouded that we never get a glimpse of its surface. But recent radar observations indicate that Venus turns on its axis only once in 243 days.

This discovery has another interesting implication, quite apart from its possible relation to the dynamo theory. If the radar observations are correct, then whenever the two planets come closest to each other on their adjoining orbits, Venus always has the same side turned toward the earth. This remarkable phenomenon can only be explained by the fact that the rotation of Venus (its turning about its own axis, not its orbit around the sun) is controlled by the gravitational pull of the earth!

The surface of Venus, as we said, remains invisible to us. Yet we must assume the existence of huge asymmetrical masses on Venus which respond to the earth's gravitational pull. Thus the earth must once have exerted a braking or decelerating effect on Venus until the two planets' mutual gravitational attraction brought about the "coupling" we observe today. Many astronomers believe that these asymmetrical masses may be huge mountains which allow the earth's gravity to "get a grip" on the planet. In 1962 Mariner II reported the existence of a "cold spot," a relatively cool zone on the otherwise glowing surface of Venus. At the time scientists thought it might be the peak of a high mountain which towered up into the cooler cloud cover of the planet. Other astronomers have theorized that the surface of

Venus might have a liquid or viscous consistency. Such a consistency would respond to the tug of the earth's gravitation, thus creating a tidal effect. The tides would explain the gravitational coupling of the two planets.

Might it be possible that the rotation of Venus is simply too slow to induce all the eddies in its liquid core to flow in the same direction? In this case, the eddies would be unable to create a unified magnetic field. In view of Venus's heavy mass, this is not a very satisfying explanation. The greater the rotating mass, the greater the tendency for it to impose uniform rotation and to inhibit eddies. This rule applies no matter how slow the planet's rotation may be.

Or does Venus lack a magnetic field because it has no moon? It might sound as if we had just plucked this notion out of thin air. But once we begin to think about the problem, this solution may not look quite so fanciful. After all, what sets the eddies in motion inside a planet's liquid core in the first place? Our dynamo theory presupposes the existence of a power source. There is only one way that the earth's liquid core could function as part of a dynamo for the production of electrical current. This liquid core would have to rotate at a slightly different speed than the solid parts of the globe. In other words, it would have to turn a little faster or a little slower than the earth's mantle. Scientists have inferred the existence of a "rotation differential" between the solid mantle and the liquid core.

Up to this point in our discussion we have been simply assuming that such differential rotation does in fact exist in the earth. We noted that much or possibly all of the earth's core is in a liquid state. But we run into difficulties as soon as we begin to examine the situation more closely. How can we take it for granted that the earth's core and its mantle rotate at different speeds? It is difficult to justify such an assumption. Yet this assumption forms the very core of the dynamo theory of the earth's magnetism. Scientists have not yet found a thoroughly satisfactory answer to this question.

A simple example taken from daily life can help us to understand the whole problem connected with differential rotation. Sometimes a few tea leaves may get into your teacup along with the tea, no matter how carefully you try to keep them out. If you twirl your tea a little, you will notice that at first the tea leaves just lie there at the bottom of the cup. Their inertia and their liquid surroundings temporarily shield them from the spinning movement. But the longer the liquid goes on turning, the faster the picture begins to change. The tea leaves start to move slowly and then pick up speed, until the entire contents of the cup are turning at the same speed.

There is a relatively simple explanation for the behavior of the tea leaves. The internal friction of the tea is very slight, as is the friction between the tea and the cup itself. All the same, this friction and the slight "viscosity" of the tea are enough to "bind" all the parts of the liquid together. Consequently the liquid and its contents soon begin to move as a single, unified object.

But what about the earth's liquid core? Why has it too not begun to rotate as fast as the solid layers of the earth? Why do they not move as one? The viscosity or inner friction of the earth's metallic core is enormous. If we assume that the earth has been rotating at the same speed for billions of years, then there is no reason why its various layers should not be moving at the same speed by now.

We have hit upon the weak point of the dynamo theory. A dynamo can function only if it contains a mechanism for generating electrical current. We cannot assume that the earth's core actually provides such a mechanism. Is there any way out of this dilemma?

Scientists have developed a supplementary hypothesis which may help to support the dynamo theory. They suggest that "thermal convection currents" may exist in the earth's core. Presumably a wide temperature variation exists between the deepest layers of the liquid core and its outer layers, which are farther from the center. Far greater pressure exists toward the center of the earth because of the weight of all the upper layers. Varying pressures cause variations in temperature; that is, the inner layers of the liquid core must be hotter than the outer layers. Hotter portions of a liquid tend to rise to the surface, whereas colder areas tend to sink.

Thus there is good cause to assume that the temperature variations in the liquid core may actually produce thermal convection currents. Unfortunately we have no way of directly proving this theory. If we accept the convection current theory, all our difficulties are at an end. If any flowing motion exists in the earth's core, then the earth's rotation has something to "catch hold on." This rotation would tend to force all the small eddies to flow in the same direction. In other words, the earth's core would act as a sort of armature, enabling the whole globe to function as a dynamo. The dynamo would then create our magnetic field.

We have now arrived at the limits of what scientists know or can reasonably assume regarding the origin of the earth's magnetic field. The dynamo theory satisfactorily explains all the basic properties of our magnetic field. In particular it explains why the magnetic poles coincide with the earth's axis of rotation. Many scientists accept the basic principles of this theory. All the same, no one has yet found a really satisfactory explanation of how the earth's core functions as the armature of the earth dynamo. Even the hypothesis of thermal whirlpools contains a number of flaws.

No scientist likes to develop supplementary hypotheses simply to "rescue" an endangered pet theory. The theory of convection currents is a copybook example of this sort of "rescue technique." We have no real proof that these currents exist. We can only say that we can find no proof that they do *not* exist. Meanwhile, their main attraction lies in the fact that the existence of thermal currents would eliminate all the snags from an otherwise flawless argument.

The second flaw in the convection current hypothesis is that probably we will never be able to prove or disprove it directly. Scientists generally frown upon assumptions which by their very nature can never be proven. The third flaw in the dynamo theory is the problem of Venus, which despite its similarity to earth lacks a magnetic field. Venus resembles the earth in size and density. Therefore it should have a liquid core composed of heavy metals. There is no clear reason why the thermal currents which presumably exist in the earth should not likewise flow inside the liquid core of Venus. We would have to construct an additional *ad hoc* hypothesis in order to explain why they do not.

The last word on this subject has not yet been spoken. Scientists are now considering another theory to explain Venus's lack of a magnetic field. One concrete difference exists between Venus and the earth which might provide an answer: The earth has a moon; Venus has not.

This theory offers the distinct advantage that scientists can directly observe all the factors it involves. The problem with the dynamo theory lies in trying to imagine how movements can develop in the earth's interior when it has been rotating at the same speed for billions of years. We shall now examine a second theory as simple as it is startling: Perhaps the whole problem does not even exist—because *the uniformity of the earth's rotation is an illusion.*

The atomic clocks of the last decade have created a new possibility. For the first time we can test the old theory of "the eternally constant motions of the stars." We can apply this test to the rotation of our own earth. We have now learned something we long suspected to be true. The astronomical time scale scientists used in the past has proved to be basically useless. One way we could state our new findings would be to say that our earth has a magnetic field because we are all a little lighter in weight during the full moon. Let us see what this statement implies.

The Universe's Time Is Out of Joint

WHAT do we really mean by the word "time"? We use it every day, and have no trouble understanding each other. But as soon as we begin to think about what time essentially is, its meaning slides mysteriously out of our grasp. At least we are in good company: St. Augustine too once commented, "But what really is time? . . . As long as no one asks me what it is, I seem to understand it perfectly well. But if I am asked to *explain* time, I am suddenly struck dumb."

Modern philosophy offers one explanation for the confusion which the concept of time inspires in us all. In the view of some twentieth-century philosophers, "time" represents one of the prerequisites of consciousness and of experience: time in itself cannot be experienced because it is a precondition of all experience. All we can experience are the various forms of "temporality." Thus when we speak of time, we ought properly to specify what kind of temporality we mean.

It might be useful to examine one of those forms of temporality which are *not* relevant to our present discussion, what the psychologists call "lived time." This is the time which we experience as being "full" or "empty," as passing "swiftly" or "slowly." The vocabulary used in measuring psychological time might strike some as rather vague or meaningless; but actually, the ways in which human beings experience time can be scientifically tested. Psychological testing reveals the factors which determine whether a person feels "time" is passing quickly or slowly. Testing can also tell us to what degree our experience of time can be altered.

This psychological, or "lived," time has its own special characteristics. Paradoxical as it may sound, only the past and the future play a role in this kind of time. Everything we think and feel has meaning only in relation to our hopes, expectations, and fears about the future, and to our past memories and experiences. The present moment shrinks to a mere illusory pinpoint.

The "objective" time of physicists and astronomers has just the opposite character. The special feature of this time consists in its *uniform flow*—in its precision and immutability. Here reality centers on the transitory present; neither past nor future actually exists.

This brief excursus on time and temporality may help to illustrate the point that a psychologist and a physicist mean different things by the word "time." Many people ignore such semantic variations and thus arrive at fallacious conclusions. For example, science fiction writers often misapply certain principles of the theory of relativity. They search out indications in the theory that time depends on the position of the observer, or that time "curves" when we approach the speed of light. On the basis of isolated, half-digested passages, authors frequently speculate on the possibility of future travel "through time." The facts in no way justify such speculations, since the science fiction writer is interpreting the word "time" quite differently from the way a physicist means it.

To discuss the problem of time more thoroughly would distract us from our basic theme—the astronomical measurement of time. But it will help us to avoid future misunderstandings if we keep in mind that the word "time" has many different meanings. During our discussion of astronomical time, we must be careful not to draw any conclusions about the nature of "time" in general.

For thousands of years astronomers took the stability of their time scale for granted. Yet this time scale has now been wholly discredited. The first revolutionary discoveries regarding astronomical time were made about one hundred years ago. Astronomers began to observe some rather uncanny things about the orbits of the moon and some of the planets. First they noticed fluctuations in the speed at which the moon was traveling around the earth. These fluctuations ("inequalities") could not be explained in terms of any known law of nature. Then similar irregularities were observed in various planetary orbits. Remarkably enough, the discovery of these new orbital inequalities did not increase the astronomers' confusion; it actually helped to put them on the right track.

For thousands of years the "eternal constancy of the stars" was proverbial in our culture. This "constancy" was fundamental to all scientific calculation: it formed the basis for the measurement of "objective" time. Physicists originally defined the meter as $1/40,000,000$ of the circumference of the earth; the meter became a basic unit of spatial measurement. In the same way, the "sidereal day" of the astronomers formed the basic unit for the measurement of time. The sidereal day, with its various subdivisions, enabled scientists to compute "universal time."

The standard used to measure universal time is the day—the time it takes the earth to rotate once on its axis. The time it takes the earth to complete one rotation is measured in relation to a particular star. Hence the more precise term for this basic unit of measure, the "sidereal" day.

Astronomers must use a star as a marker by which to measure the earth's rotation, for how else could they determine when one rotation is complete, since the planet they are clocking rotates in empty space? Our sun, the moon, and the other planets move much too fast relative to the earth to be used as markers. To be sure, the "fixed" stars also move. But they are almost all so far away that even the most delicate modern instruments cannot detect their movements. Therefore stars can serve as accurate markers. In order to measure the duration of the earth's rotation, an astronomer chooses an appropriate star. He then observes the star through a special telescope which moves only along a north-south line (along a "meridian"). As the earth turns, the star moves from east to west across the sky. The astronomer tries to determine the exact moment when the star crosses his telescopic sight. Then he measures the time which elapses until the same star once again passes through the line he has marked on his instrument. At this instant the earth has completed one rotation.

The time which elapses between two such "sidereal transits" is called a sidereal day. Astronomers have continued to refine their techniques of measurement. Automatic photoelectric methods can now be used to determine the exact moment of a star's passage. Various errors in measurement have been analyzed and eliminated. (For example, telescopes sometimes alter their position very slightly as a result of temperature changes.) These improvements have made it possible to determine the duration of a sidereal day to within 1/1,000 second.

The sidereal day forms the basic yardstick for the measurement of all celestial phenomena. The subdivisions of this basic time unit are the hours, minutes, and seconds. Special instruments divide the day into segments which are as nearly equal as possible. One such instrument is the clock. In order to measure the motions of planets, moons, and comets in our solar system, two things were necessary: precision clocks and increased accuracy in measuring the sidereal transit.

For centuries the history of astronomy has basically consisted in the gradual improvement of these two essential factors.

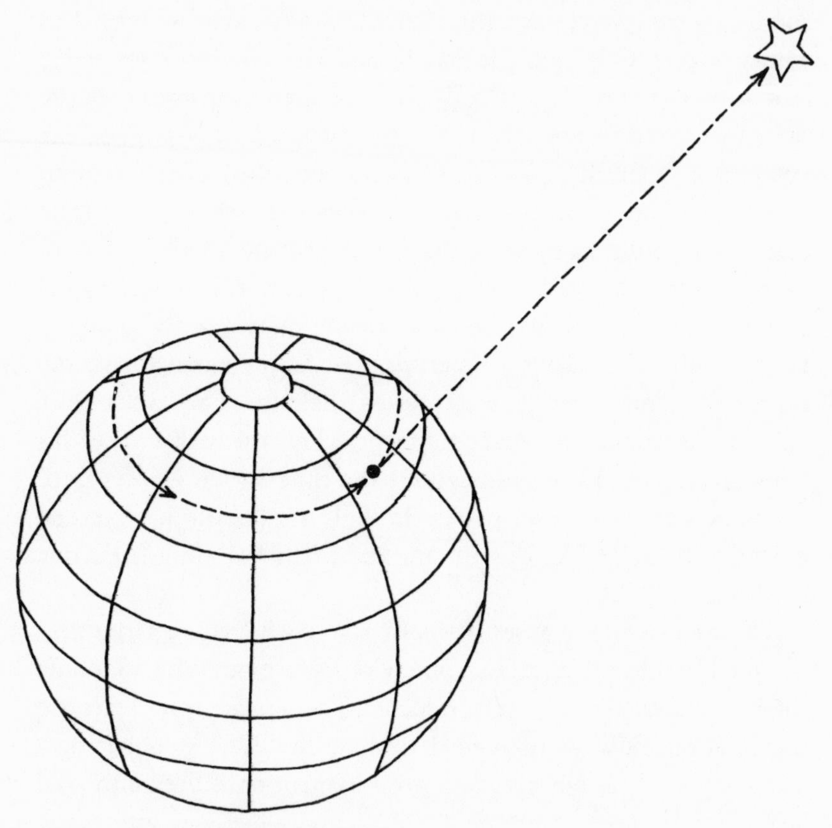

How scientists determine the length of a sidereal day.

Scientists have continually struggled to achieve a more precise division of hours, minutes, seconds, and even fractions of seconds.

Four thousand years ago the Egyptians were patiently observing the heavens. They learned that in one year (the time it takes the earth to complete one revolution around the sun) the earth has completed 365.25 rotations on its axis. By modern standards this figure is very imprecise; but it is still basically correct. It comes within 1/100 day (about a quarter of an hour) of the actual length of an average year. In antiquity no reliable instruments existed for observing the heavens. The Egyptians had to trace the course of heavenly bodies with the naked eye. Yet their observations were surprisingly accurate. No one improved on their measurement of the sidereal day until around A.D. 1600, after the invention of the telescope.

After the telescope came into use, scientists observed that the relationship between the day and the year sometimes seemed to fluctuate. At times the earth appeared to have rotated farther in one year than in another. These variations could not have resulted simply from the imprecision of seventeenth-century scientific instruments. Instead scientists tended to attribute the fluctuations in measurement to mistakes on the part of the observer.

Until the seventeenth century, scientists were unable to divide the day into smaller units with any degree of precision. Then two seventeenth-century inventions increased the precision of scientific measurement. The first was the pendulum; the second, the portable spring-driven clock with its freely suspended balance wheel. Because of the unprecedented regularity of their movements, the pendulum and the balance wheel proved ideal instruments for subdividing time segments.

The clocks which had been in common use up to this time fulfilled their everyday functions quite efficiently; but they were useless for purposes of astronomical measurement. Despite all the efforts of the clockmakers, their clocks had tended to gain or lose a quarter of an hour daily. These clocks were all based on the same principle. The clockmaker measured out a specific quantity of a suitable material, which was then made to flow at an even rate. Then in various ways further divisions were made in the flowing material. Generally water or sand was used; but the renowned Danish astronomer Tycho Brahe (1546–1601) experimented with quicksilver. Actually the flow of these water or sand clocks was quite irregular. The pendulum and the balance wheel, on the other hand, at once achieved almost perfect accuracy; they gained or lost as little as a few seconds a day.

Important as these innovations were, they failed to satisfy the strict requirements of astronomers and sailors. In the long run it was fortunate that they did not. Sailors at once recognized the significance of spring-driven clocks for navigation on the open sea. We have already noted that the time it takes the earth to rotate once on its axis must be determined in relation to a star. The opposite also holds true: It is possible to calculate where a particular star will be at a particular time in relation to a specific vantage point on the earth. A navigator on the open sea can measure the location of the stars with a suitable instrument (the sextant). In this way he can find out where he is. Of course, the earth's rotation constantly alters the position of the stars. Therefore the navigator must take care to measure his guiding star at just the right moment. In other words, he is dependent on the accuracy of his ship's clock. If his clock is inaccurate, then the navigator cannot determine his location or plot an accurate course.

The spring-driven clock represented a major advance over earlier timekeeping devices. Even by today's standards, a clock which gains or loses only a few seconds a day qualifies as a remarkable technical achievement. Few of us own watches which keep such good time. Yet a seafarer may spend whole weeks or months on a sailing ship; a loss of even a few seconds a day soon adds up to a sum which throws off all his calculations. Let us assume that he has a clock on board that gains only three seconds a day. Forty days pass (in the eighteenth century, an Atlantic crossing would have taken at least this long). By now a discrepancy of 120 seconds exists between his clock time (on which he depends for locating various stars) and true sidereal time. During these 120 seconds the earth has been turning. The star the navigator tries to sight has already traveled a good stretch from where he ought to have found it.

Particular stars are found at specific locations only at specified times. Navigational charts relate these times to areas on hydrographic charts. If a star is sighted too soon or too late, the error automatically leads to a false determination of the ship's location. However, an error in navigation may have less serious consequences if a ship is sailing on a route that lies far to the north or south of the equator. Near the poles the earth's surface moves more slowly from west to east than at the equator. (At the poles themselves the earth's surface is still turning, but no real lateral movement occurs.) Thus the stars in the polar regions seem to move more slowly across the sky than stars at the equator. Areas along the equator have to travel nearly 25,000 miles every twenty-four hours. That is, they must travel at the speed of a modern jet plane—well over 1,000 miles per hour.

Thus the 120-second time discrepancy suffered by our seafarer might have had various consequences for his ship. These consequences would have depended on the degree of latitude at which the ship was traveling. If it was near the equator, the navigational error could have thrown it more than thirty-two miles off course. If the ship had missed its intended port and arrived on a strange coast, the navigator might not even have known whether to search for the port to the north or the south.

One can easily imagine the enthusiasm which greeted the invention of the spring-driven clock. It delighted the admiralties of all the great seafaring powers of the seventeenth and eighteenth centuries. At last a method existed for determining a ship's location after weeks of cruising on the open sea. After this the only problem lay in increasing the accuracy of the "chronometers," as seafaring men had christened the new timepieces.

The admiralties' interest in developing more precise timepieces had some fortunate consequences for astronomy as well as for navigation. It often happens that science indirectly benefits from a political or commercial enterprise. (And sometimes the results are not quite so harmless as they were in this case.) Suddenly seventeenth-century scientists were being offered money and prizes for research. Rival navies began to compete in the fabrication of superior chronometers.

Even the great Newton at once seized the opportunity to use navy funds to improve the quality of British timepieces. At his instigation, the English government in 1714 offered researchers a prize of 20,000 English pounds (in those days a fantastic sum). This fortune would fall to the first man to make a chronometer capable of traveling from Europe to America and back without gaining or losing more than one minute of time. The clock was to leave London for an American port and then return—an estimated journey of some 120 to 160 days. At no time during this journey could the clock be reset.

Even today this would be a hard nut to crack; few present-day watches could pass such a test. Newton himself did not live to see the solution to the problem he had posed; he died in 1727. The British government did not have to dig into its pockets until almost half a century after the prize had first been offered. The lucky winner was John Harrison, a skilled shipwright from the county of York, whose passion for building and inventing had made him into a fine clockmaker. For most of his life Harrison worked at trying to solve the seemingly insoluble task set by Newton and the British navy. He came near to failing altogether. Harrison was sixty-eight years old in 1761, when he finally realized his goal. A Harrison chronometer traveled with a ship on its 151-day journey from London to Jamaica and back. Upon its return to London, the chronometer differed from London time by only fifty-six seconds.

Let us now make a leap of two centuries and compare how the same problem is being dealt with today. During the last decade people have been able to travel from Europe to America and back without losing more than $1/1{,}000{,}000$ second. This figure represents a ten-millionfold increase in precision since the year 1761. This incredible increase in precision resulted from the most recent innovations in chronometry—the so-called atomic clocks. (These clocks will be discussed in the following chapter.) The synchronization between the two continents was effected by two means. First, time signals were exchanged through the artificial Telstar satellite. Second, scientists used the same basic method that had been employed with Harrison's chronometer some two centuries before. An atomic clock was transported by plane from Europe to the United States. Here it was used to reset an American atomic clock. Finally it was flown back to Europe, where it was corrected by a third atomic clock which had never left the spot. (In the next chapter we will discuss the usefulness of synchronizing various regions of the earth so that their timepieces differ by only a few millionths of a second.)

Thanks to the talent of John Harrison and of later inventors all over the world, the precision of the chronometer rapidly increased. By the nineteenth century, chronometers had been developed that lost or gained only fractions of a second a day. The far from purely scientific zeal of the admiralties had contributed to this effort. At the turn of this century, some pendulum clocks were housed in special casings which protected them from shocks and from changes in temperature and humidity. These clocks gained or lost only a few one-hundredths of a second each day.

The precision of chronometers in the narrow sense of the word never really increased beyond this point. A new invention had made the seafaring peoples quite content with the chronometers they already possessed. The invention was that of wireless telegraphy. Telegraphy made it possible to communicate with a ship at sea. In the past, ships had been totally isolated for long periods. The small inaccuracies in their chronometers had accumulated steadily day by day, causing errors in navigation. But henceforth the true time could be relayed by telegraph, so that chronometers could be reset daily.

Under these new conditions, chronometers no longer had to be exquisitely accurate; they could lose or gain one second each day without throwing off the navigator's calculations. Even at the equator, a time deviation of one second would throw the ship only a quarter mile off course. In any case, the sextant itself had never been able to determine star elevations with absolute accuracy. Therefore the precision of the chronometer had never been turned to full account.

The admiralties were now content with their chronometers. One might have thought that the astronomers would have been content too. They had invested centuries of effort in the precise measurement of their basic unit, the "day." The sophisticated pendulum clocks made it possible to break down this day into a number of highly exact tiny segments. Planetary orbits could now be measured with unprecedented accuracy. After all, in the eyes of astronomers the principal value of precision timepieces lay in the possibility of defining the orbits of various heavenly bodies. A knowledge of these orbits would help astronomers to explain other processes taking place in our solar system.

But when scientists used the new instruments to measure the solar system, they suffered quite a shock. They had simply assumed that they would now be able to calculate orbits more accurately. Just the opposite occurred. The new instruments seemed useless for purposes of astronomical measurement. Whenever astronomers tried to use the new timepieces, they became dizzy: The whole solar system seemed to be pulsating in a sort of rhythmical pattern. This pulsation violated all the laws of celestial mechanics which astronomers had taken for granted since Kepler and Newton. What was happening to the world?

Let us look at the startling sight that confronted astronomers when they began to use their new instruments. Our diagram shows the speeds at which the moon was orbiting the earth during the period between 1750 and about 1920. The horizontal line represents the normal estimated speed of the moon. The curved line marks the direction and degree of deviation from this norm. The numbers along the horizontal line refer to the date.

We can see that in 1750 the moon was traveling more slowly than it theoretically ought to have been. Yet its speed continued to decrease until the time of the French Revolution, when the degree of deviation was almost twice what it had been in 1750. From then on the speed seemed to slowly increase again. It took the moon almost seventy years to catch up to the speed at which it ought theoretically to have been traveling all along. By 1816 it had succeeded in doing so; yet its speed continued to increase. By the turn of the century this speed surpassed the "normal" speed by about the same margin as it had fallen short of it before. At this point the braking effect once more came into play. The moon's speed leveled off for a few years around 1920 before it began to rise again.

In other words, the moon was literally behaving "impossibly." What force in the universe could have caused the bizarre phenomenon scientists now referred to as "periodic inequalities of the lunar orbit"? What mysterious power was capable of accelerating the speed of our moon for a hundred years, only to abruptly slow it down again?

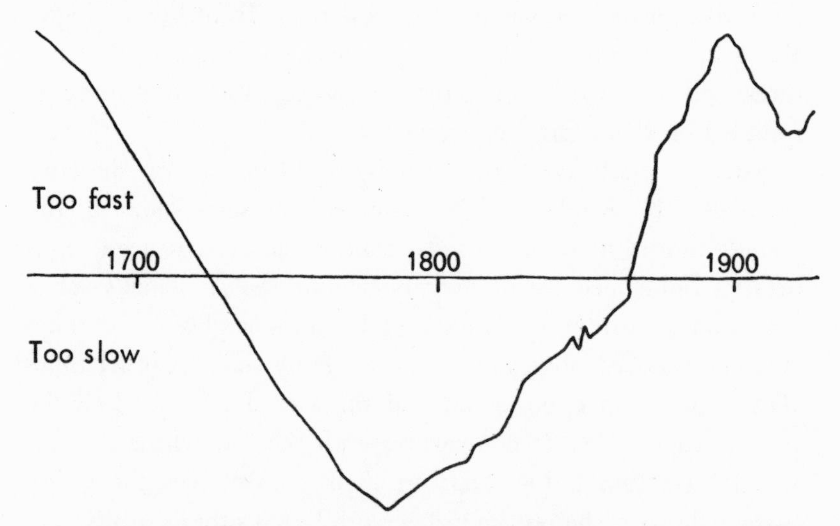

Fluctuations in the speed of the moon's orbit during the last few centuries. Does the whole solar system obey the same rhythmical pulsation?

To be sure, disturbances in the orbits of various heavenly bodies were a commonplace occurrence. Astronomers wanted to study instances of such "perturbations." This was one reason why they persisted in measuring the orbits of planets, moons, and comets over and over with the greatest possible precision. Their computations enabled them to determine the orbits of newly discovered planets and comets. But apart from the study of newly discovered heavenly bodies, astronomers had another reason for devoting so much time to orbital calculations.

Kepler had established the laws of celestial mechanics. Even now no one seriously dreams of challenging these laws. Since the beginning of the seventeenth century, every astronomer has known that all planetary orbits trace an ellipse which has the sun as its focus. He has also known that the orbital speed of all heavenly bodies depends on their distance from the sun. Even today no one can say why this should be so. The heavenly bodies obey laws which can seemingly be represented in terms of simple and elegant mathematical equations. It is a mystery why any correspondence should exist between the motions of the stars and the mathematical laws derived from human logic. Yet a correspondence does exist between mathematical structures and the world of physical nature; and all science is founded on this central mystery.

Astronomers knew that no doubt could be cast on the laws discovered by Kepler and Newton. All the same, none of the planets moves in ideal solitude around the sun. Instead, each orbit is influenced by its neighboring orbits. A planet's speed may vary according to its distance from its neighbors. At times two planets may move along almost side by side. At other times they may be on opposite sides of the sun. To be sure, all the planets appear tiny in comparison with the sun whose gravitational attraction holds them on their course. Yet these tiny planets do have the power to "perturb" each other's orbits.

Thus all the planets deviate somewhat from Kepler's ideal

ellipse. A planet could maintain an ideal ellipse only if it were alone in the universe except for the sun.

After a while, locating these minimal orbital perturbations became a sort of competitive sport among astronomers. For one thing, the complex mathematical problems posed by the perturbations challenged their ambition. Mathematically speaking, the problems posed by "multibody theory" (involving mutual influences between two or more heavenly bodies) represented virgin territory. Moreover, knowledge of orbital perturbations supplied astronomers with much valuable information. For example, this knowledge enabled astronomers to draw some conclusions about the mass and physical composition of heavenly bodies. These data could not have been obtained in any other way.

The entire scientific world, not to mention the general public, were greatly impressed by the discovery of the planet Neptune in 1846. Neptune was literally discovered at a desk.

Until 1781 everyone believed that only six planets (including the earth) were orbiting our sun. Then the rather obscure astronomer Herschel discovered Uranus and became a world-famous figure overnight. The orbit of the new planet was eagerly studied in observatories all over the world. This orbit soon began driving astronomers to distraction. They were unable to establish its dimensions conclusively. Many times all the observers would agree on these dimensions; then they confidently waited for Uranus to turn up where it was expected. But each time they were disappointed. Uranus never appeared at just the spot prescribed by mathematical law. True, it was never very far from where it should have been—never more than two minutes of arc. (By way of comparison, the full moon has an apparent diameter of a little more than thirty minutes of arc.) All the same, by this time astronomical measurements were far too sophisticated to have erred by two whole minutes of arc. Such a discrepancy violated Kepler's laws. It simply could not be.

At this point, one possible alternative was to invent some sort of hypothesis to explain away the discrepancy; perhaps Kepler's laws were no longer in full force at such an enormous distance from the sun. This hypothesis was not very convincing. The only other possibility seemed to be that some huge mass near Uranus was "perturbing" its orbit. The orbit of Saturn (Uranus's neighbor toward the sun) showed no perturbations which could not be explained through its proximity to Jupiter on the one side and Uranus on the other. Therefore the mass perturbing Uranus had to lie *outside* the orbit of Uranus, and not among the known planets. In other words, Uranus was *not* the outermost planet; the solar system might be far larger than anyone had imagined. At least one more large planet had to exist beyond the newly discovered Uranus. There began a feverish search for the "transuranian" planet.

This search was carried out at a desk. No one would have known where in the sky to look. Presumably the light from the unknown planet was very weak. It could never have been found by just aiming a telescope at random. There seemed to be only one method which might work. First, scientists calculated the perturbations in the orbit of Uranus. The resulting figures were used to compute the speed and the location of the unknown planet. Only an astronomer can imagine the difficulties involved in this task. Yet in 1846 these difficulties were solved simultaneously by two scientists totally unacquainted with each other's work. They were the Englishman John Adams and the Frenchman Jean Joseph Leverrier.

Adams achieved his results several months before Leverrier; but unfortunately he did nothing more than report them to another astronomer at the Cambridge Observatory. The observatory records indicate that probably this Cambridge astronomer actually saw the new planet twice in the days following his conversation with Adams. In any case, Adams neglected to publish his findings. The Frenchman came away with all the laurels. On August 31, 1846, Leverrier published the results of his computations. At the same time he wrote the Berlin astronomer Galle, requesting that Galle look for the new planet at the place Leverrier suggested. Galle found the hitherto unknown member of our solar system at almost exactly the indicated spot. Leverrier's feat was greatly admired. Scientists were delighted at this vindication of Kepler's laws. They also saw what benefits could be reaped from the increased precision in astronomical computation.

Almost one hundred years later the same method was used to detect another planet. More refined techniques of observation played their part in this second discovery. In 1930 the American astronomer Clyde Tombaugh located a tiny point of light on a photographic plate: It was Pluto, a small planet whose orbit lay beyond that of Neptune. The computations which enabled Tombaugh to make this discovery were actually performed by another American, Percival Lowell. Lowell's calculations were based on perturbations of the orbit of Neptune. Even though these perturbations amounted to only a few seconds of arc, they gave Lowell the information he needed. Some astronomers now believe that Pluto may not be the outlying planet in our solar system. Pluto's orbit also exhibits small perturbations which have not yet been satisfactorily explained. Probably we will not have a final answer to this question for several decades. Pluto's mean distance from the sun is almost 4 billion miles. Consequently it moves very slowly along its orbit. Astronomers began to observe it in 1930. But they have not yet had an opportunity to observe the perturbations of its orbit for long enough to draw any real conclusions.

Thus orbital perturbations offered astronomers a constant challenge. Then the astronomers began to reexamine the orbit of the moon, using the newest timepieces. The inequalities of the lunar orbit turned out to be quite different from the perturbations observed in connection with the newly discovered planets. The moon was not even 239,000 miles from the earth; clearly no unknown heavenly body could be causing its abnormal behavior. Where was the flaw in the astronomers' reasoning? How was it possible that the moon should first decrease its speed, and then all of a sudden increase it again? The moon was pulsating in what might be interpreted as a sort of symmetrical rhythm. Had scientists encountered a hitherto unknown power of nature?

Astronomers fell prey to disquieting suspicions. Perhaps they were looking at the whole problem backward. Perhaps the moon was behaving normally, and only their measurements were flawed. The astronomers suddenly recalled the difficulties they had recently been having in establishing the exact relationship between the day and the year. We have already noted that in some years the earth seemed to have rotated a little farther than expected—not exactly 365.25636 times, but slightly more.

The astronomers became suspicious of everything. They began to reexamine the orbits of our neighboring planets. These planets traveled much more swiftly than the moon, and therefore any irregularity in their orbits would quickly become apparent. The worst fears of astronomical mathematicians were confirmed: Assumptions basic to astronomy for thousands of years proved to be an illusion. All the planets they investigated exhibited the same orbital perturbations as the moon. And most startling of all, these perturbations appeared to be completely synchronized! The planets seemed to speed up or slow down whenever the moon did. Speeds throughout the solar system displayed a rhythmic pulsation. True, the pulsation was slight. But all the planets were keeping perfect time!

There was only one possible explanation for this phenomenon. Scientists had not discovered a new force of nature, but simply a phantom. The rhythmical pulsations of the solar system were unreal. They were an illusion created by rhythmical deviations in the unit used to measure them, the sidereal day. Incredible as this conclusion might seem, the earth's rotation was clearly not as uniform as astronomers had always assumed.

An analogy can help us to picture what was happening here.

Suppose that a sports club has sent a team of runners to a track meet. All the runners have been carefully trained, and the captain knows the "personal time" of each team member. On the first day of the competition the captain discovers that each of his men has made better time than ever before; in fact, they have all increased their speed by exactly the same amount. For a moment the captain might be delighted. But think how disconcerted he would be when he learned that all the other competing teams have improved to exactly the same degree! If such an incident really should occur, probably everyone would guess right away that something was wrong with the timekeeper's stopwatch.

The timepiece clocking the speed of these runners was running a little slow. For example, if a stopwatch registers 10.0 seconds when actually 11.0 seconds have gone by, then a merely "good" sprinter would suddenly seem to be setting world's records. Just the opposite would occur if the watch were running fast. It might register 12.0 seconds instead of the actual 11.0. In this case the fairly good sprinter would be magically transformed into a merely average one.

The same phenomenon had caused the rhythmical pulsations of speed inside our solar system. The pulsations were merely *apparent*. The clock being used to measure them sometimes ran slow or fast. This clock consisted of the sidereal day with its various subdivisions. The basic unit of astronomical measurement had been the earth's rotation. Increasingly accurate timepieces had divided this basic unit into smaller and smaller subdivisions. But astronomers had assumed all along that sidereal days were always exactly equal in length. After all, they had no reason to believe that the earth did not rotate with absolute regularity. This assumption now proved false. No wonder that astronomers were dismayed!

This unexpected revelation posed two grave problems for astronomers. Their first problem was that of finding a basic unit of measurement to replace the sidereal day. "But if the salt have lost his savour, wherewith shall it be salted?" The cosmic clock which astronomers had used in the past had turned out to be unreliable. But what other basic unit could they use to measure celestial phenomena?

The second problem confronting astronomers was that of discovering the cause of the irregularities in the earth's rotation. The earth was turning freely in empty space. What could possibly be causing it to move in fits and starts? Eventually a single invention provided the solution to both problems—the atomic clock.

Cosmic Pirouette

WHENEVER the autumn leaves are falling from the trees, the earth turns a little faster than before. Thus in autumn a day is a little shorter than the "normal" twenty-four hours. The discrepancy is slight, but it has been clearly measured; autumn days have lost some .06 second. Each spring the days increase their length by exactly the same amount. Many such inequalities disrupt the supposedly uniform rotation of our earth. These inequalities were discovered as soon as new clocks had been invented to replace the unreliable "natural" timepieces. The new clocks showed that the proverbial constancy of our planet was not all it had been cracked up to be.

Quartz clocks were invented in 1929. Electrical current shot through a quartz crystal causes the crystal to vibrate with very swift but uniform motions. Uniform vibrations form the basis for all measurement of time, including the pendulum clocks and the spring-driven clocks with their oscillating balance wheels. Some difficult technical problems had to be solved before a vibrating quartz crystal could be used as a clock. Scientists had to find a method of recording or counting the vibrations; but this method could not interfere with the vibrations themselves. Eventually researchers overcame all the technical obstacles. It became possible to keep time without losing or gaining more than $1/_{1,000,000}$ second per day.

The quartz clocks represented a major advance over earlier forms of timekeeping. But they had one disadvantage which is common to all clocks of this type. The constant electrical vibrations rapidly altered the mechanical properties of the quartz crystal. The frequency at which the crystal vibrated depended on its original mechanical properties. In other words, at first the quartz clocks achieved an unprecedented degree of precision. But this miraculous precision lasted at most a few months. Thus the quartz clocks were not suitable for the purpose of examining the earth's rotation, which must be measured over the course of an entire year.

Ten years ago scientists developed new clocks capable of measuring our standard of astronomical time, the earth's rotation. Like the sidereal day, these revolutionary clocks are based on a process of nature—but not on the unreliable motion of a heavenly body. They are based on the vibrations of atoms in particular elements. We need not concern ourselves here with the actual construction of the clocks, or with the problems of recording high-frequency vibrations. We need only note the fact that the atoms of each element have their own individual rate of vibration. Therefore the precision of such an "atomic clock" seems to be unimpeachable.

It would be virtually impossible to describe the incomparable precision of this new generation of clocks. Expressed in figures, this precision amounts to 10^{13}, or a ratio of one to ten trillion. That is, the clocks are accurate to within one part in

10,000,000,000,000. All this sounds very impressive, but is not especially helpful to nonmathematicians. A mental game may help us to picture the accuracy of the new time standard more clearly. Suppose that two atomic clocks were built in the year 0 and were perfectly synchronized when they began to run. After this they were simply left alone. Today, almost 2,000 years later, the time kept by these two clocks would differ at most by .001 second!

But what is the sense of all this unearthly precision? It might delight a technician obsessed with exactness for its own sake if he could dissect a millionth of a second into 10,000 additional perfectly equal segments. But the layman is bound to wonder what could possibly be measured on a scale of millionths of seconds. Actually, scientists are not attempting to measure such minute phenomena. The importance of the atomic clocks lies in their ability to synchronize various events and to measure the constancy of various processes.

In a previous chapter we noted that Europe and America were synchronized by means of three atomic clocks, one of which was transported back and forth by plane. The two continents were synchronized to within $1/1{,}000{,}000$ second. Why was it important to expend all this effort? A number of answers might be given to this question. For one thing, such an exact synchronization of the two continents made it possible to establish a completely new aircraft guidance system. Nowadays it is no longer the admiralties that provide funds to perfect techniques of measurement; these funds are now furnished by the Air Force. A guidance system based on time comparison of radio signals is independent of the weather, unaffected by atmospheric or man-made disturbances.

The new guidance system is based on the following principle: A number of transmitters are distributed at various locations over a wide area. At brief intervals all the transmitters simultaneously emit a short signal. These signals are received by the airplane. Of course the signals do not all reach the plane at the same time, since the transmitters are located at varying distances away. Despite its speed, a radio signal always requires a certain amount of time to reach the plane. A quartz clock on board the plane records the arrival times of the various signals. Each signal has its own frequency, which identifies its point of origin. The time intervals are fed into a computer which converts them into corresponding measurements of distance. The navigator then records these distances on a chart showing the locations of all the transmitters. In this way he is able to determine the present location of the airplane.

One obvious precondition of this system is the precise synchronization of the transmitters emitting the time signals. Unless the signals are emitted *simultaneously*, the time intervals cannot be properly measured on board the plane. Radio signals travel at 186,000 miles per second. At this speed, a time difference of 1/1,000,000 second corresponds to a distance of about 985 feet. If all the transmitters were synchronized to within "only" 1/1,000,000 second, the plane would end up way off course. To insure synchronicity, an atomic clock is transported from one transmitter to the next; this clock is used to reset the atomic clocks at all the stations. In this way the "simultaneity" of the signals can be guaranteed to within a few hundred-millionths of a second. This degree of synchronicity would result in errors of only a little over one hundred feet. (Of course, unavoidable errors in measurement or computation always further reduce the level of precision.) As technicians increase the "simultaneity" of the transmitters, the range of possible error continues to decrease. In time, errors will be confined to fractions of an inch. One day a fully automatic robot pilot system can be established. Computer-controlled landings will take place safely at night and in thick fog.

This principle is already being turned to military use. For example, submarines can use it to determine their positions while they are still submerged. The submarine crew measure the time intervals between signals sent out simultaneously from different transmitters. A receiver buoy on the surface is used to pick up the signals. This method has the advantage that the submarines themselves need not betray their position by sending out their own sounding signals. Precise synchronization is equally essential in the field of astronautics; for example, it can be used to guide space probes. It is reassuring to know that once this whole process can be made efficient and economical enough, it will benefit even lowly civilians: It will eventually be used in automatic landing systems to guide international air traffic.

Apart from the synchronization of timepieces, the new atomic clocks also perform a second function. They make it possible to test the constancy of various astronomical processes more precisely than ever before. These clocks can help astronomers answer some of the questions which have been plaguing them.

At the beginning of this chapter we noted that the earth rotates a little faster in the autumn and a little slower in the spring (a deviation of only .06 second every twenty-four hours). Without the atomic clocks we would still be unaware of these seasonal inequalities in the rotation of our globe. Astronomers have not yet fully explained the phenomenon. The periodic irregularities are so slight that we never feel their effects; nor could astronomers have detected them by any form of measurement other than the atomic clock. Such slight inequalities would have to accumulate for a long period in order to register on an "ordinary," or nonatomic, clock. But the inequalities do not have time to accumulate, since they are balanced out again in the course of a year.

At first one might wonder whether these periodic inequalities in the speed of the earth's rotation actually exist. After all, the moon too had seemed to vary its speed as it traveled around the earth; yet these variations had proved an illusion. The apparent alterations in speed had really been caused by a flaw in our unit of measurement. Could not these strange periodic inequalities in the earth's rotation likewise be caused by an error of measurement? In other words, can we be absolutely certain that our atomic time measure is really constant?

The answer must be no. We have no way of testing the constancy of atomic frequencies unless we one day find an even more precise standard to measure them by. At this stage, it appears theoretically impossible to discover a more precise standard. Besides, finding a new form of measure would not solve our problem, since we would need yet another measure to prove the validity of that one, and so on. In any case, we cannot doubt the reality of the periodic fluctuations in the speed of the earth. True, many things about these fluctuations have not yet been explained. But a basic relationship clearly exists between the inequalities and certain seasonal changes taking place on the earth.

We cannot know if the atomic time standard is truly constant. Scientific logic and experience argue that it is. All the same, we can only measure temporal processes by comparing them with another process which we assume to be unfolding at a uniform rate. The earth's rotation was the first such "uniform" process used by the astronomers. Eventually even the lowly pendulum clock revealed the unreliability of this standard. Now we use atomic clocks to test the classical standard, the earth's rotation. But we have no way of testing the atomic clock. It is pure speculation to suggest that at some point in the history of the universe, the vibrations of certain atomic frequencies might increase or decrease in number. No one could prove or refute such a speculation. But if such a thing ever happened, it would invisibly flaw all the formulas with which we tried to describe the universe.

To measure is to compare. Any measurement stands or falls on the accuracy of its basic unit of comparison. Let us suppose that a supernatural demon had the power to shrink or to enlarge the whole universe—everything from subatomic particles up to intergalactic space. The demon could shrink or expand all this thousands or even millions of times. In such a case, we would notice absolutely nothing, since both we and all our standards of measurement would participate in the change. If the demon should decide to slow down or accelerate all the processes taking place in the universe, or even to stop the world altogether for a while, we would never notice the event.

We must keep in mind that all these reflections and speculations have nothing to do with time itself. We are simply discussing the basic factors confronting any scientist when he tries to define and measure "objective" time. Our reflections are intended to point out that we have no choice but to deal with time as a purely pragmatic concept. As scientists nowadays are fond of saying, we proceed on a purely "operational" basis. The actual "operation" or procedure a scientist uses to measure something defines the nature of what he is measuring (in this case, time). He does not attempt to speculate over what time really "is," or how time could exist in this abstract form at all. Before 1965, a second was officially defined as 1/31,556,925.9747 of a year. Since 1965, a second has equaled the time it takes a cesium atom to vibrate back and forth exactly 9,192,631,770.0 times. This official definition of the second was issued by the International Union of Weights and Measures at a general conference held in Paris in 1964.

During the past ten years atomic clocks have enabled scientists to precisely measure inequalities in the earth's rotation. These inequalities have been traced to a variety of causes. Scientists had been astonished to discover that the earth does not rotate with absolute uniformity. Imagine their surprise when, with the aid of the new clocks, they learned that actually a whole series of factors were disrupting the earth's pirouette in empty space! They have not yet had time to track down all these factors. In the previous chapter we noted those periodic rhythmic inequalities in the earth's rotation which could sometimes extend over whole centuries. Clearly, our scientists could not in one decade have analyzed a pattern which takes centuries to unfold. Thus we do not as yet have a plausible explanation for these lengthy rhythmical fluctuations.

The brief "seasonal" inequalities which balance each other out in the course of a year are quite another matter. Astronomers have been able to observe this cycle and to form some theory about it. Our planet actually does perform a cosmic pirouette! The earth increases and decreases its speed very much as an ice skater alters her speed while performing a pirouette on the ice. As she turns, she sometimes holds her arms close to her body, so that she turns more quickly; then again she spreads her arms wide, and the turning motion slows down. She can perform the same maneuvers again and again. In terms of the laws of physics, the spinning impulse with which she began her pirouette can be prolonged indefinitely. This impulse never gets "used up": the pirouette need not cease until the skates have worn down the ice so that her skates no longer slide on it properly. The skater can continue to convert the original spinning impulse into two different movements. She may rotate swiftly, with her arms close to her body; in this case she will cover only a little ground at each turn. Or she may move more slowly, her arms and hands extended; in this case some parts of her body will traverse a greater distance.

The laws of mechanics apply to all objects, whatever their nature. Therefore everything we have noted about the ice skater also holds true for the earth. The initial spinning impulse is not lost during the earth's seasonal fluctuations in speed. Thus the earth too creates a pirouette effect. But where are our earth's "arms"—those parts it extends into space during the spring, in order to slow itself down? And what is it that the earth "holds close to its body" when it speeds up in the fall?

Probably it is *water* which moves in a seasonal rhythm between the earth and its atmosphere. In the spring the sun's rays increase in intensity, thus warming the surface of the earth. The upper layers of the earth's crust dry out. That is, a great quantity of moisture passes into the atmosphere in the form of water vapor.

Every spring and summer, billions of tons of water rise into our atmosphere. They may rise as high as 3,500 feet. These quantities of water function as the "arms" which the earth extends into space, slowing down its rotation. The water vapor rises only a few hundreds or thousands of feet above the earth; therefore the water loss does not appreciably alter basic conditions on our planet. The loss of water vapor slows down the earth by at most .06 second per day. In autumn, the effect is reversed; in a short time all the water ballast has been rained down on the earth again.

Plausible as this explanation may sound, does it not contain a flaw? The seasons unfold at different times in the northern and southern hemispheres. When it is spring in the north, autumn has come to Australia and South Africa. Strange as it may seem to northern peoples, November has the same romantic attributes for people in southern climes as May has for them in the north. The two seasonal fluctuations we have just described, which apparently control the earth's pirouette, would seem to counterbalance each other. Whenever it is spring in the north, water is evaporating; but at the same time autumn rains are falling in the south. Logically, the two effects ought simply to cancel each other out. If true, this factor would refute the entire "pirouette" theory as an explanation of the seasonal inequalities of the earth's rotation. But we need only look at a globe or map of the world in order to see why this "balancing" effect does not really take place. The land masses in which the widest temperature variations occur are irregularly distributed on our globe. The continents of the northern hemisphere are much larger than those of the southern hemisphere. Therefore the seasonal changes occurring in the north completely outweigh the effects of those in the south. The northern seasons determine the pirouette effect.

It has not yet been established whether water movements between the earth's crust and its atmosphere are solely responsible for the pirouette phenomenon. Other seasonal factors may also contribute to this phenomenon. At the beginning of this chapter we noted that the earth turns a little more slowly in the autumn when the leaves are falling from the trees. Leaves are very light, and trees are not very tall. But there are a great many trees on the earth, and innumerable leaves, all of which fall to the ground at more or less the same time. Even though the leaves fall such a short distance, they redistribute the mass of the earth; they make the earth more compact, as the ice skater makes her body more compact by holding her arms to her sides. Some well-known scientists believe that even the falling leaves may contribute to the periodically quickening tempo of our cosmic pirouette.

The Moon-Brake

THERE is an exception to every rule." This sentence represents a contradiction in terms, for it too is a rule and must therefore have an exception. Paradoxical though it may be, the rule certainly applies to that youthful branch of geophysics which deals with inequalities in the earth's rotation. We have established the rule that brief inequalities of the earth's rotation are more readily analyzed than those whose pattern extends over centuries. We are about to encounter two exceptions to this rule.

A number of slight fluctuations in the speed of the earth's rotation can scarcely be detected. We have no way of predicting when these abrupt accelerations and decelerations will occur. If they were more powerful, the earth would constantly be suffering major disasters. We have only to imagine the sort of thing that happens when the emergency brake is suddenly pulled in a crowded train. Think what would happen to the earth if it were to abruptly change speed! Houses, people, trees, loose surface soil, even whole mountains would be torn up by the roots and hurled toward the east as if by a magic hand. How fortunate for us that these brief fluctuations in speed do so little damage! At least they have not done so up to now. If past fluctuations had triggered worldwide disasters, these events would have produced major alterations in the earth's crust. Our present geological research techniques would have enabled us to detect these changes even after the passage of hundreds of millions of years. But geologists have discovered no such evidence.

As yet we have no idea what causes these abrupt fluctuations in the speed of the earth's rotation; thus we could not control them even if we wanted to. Some geophysicists suspect that the fluctuations may be caused by an irregular shifting of solid masses in the liquid layers of the earth's core. This shifting movement might momentarily displace the earth's center of gravity. This hypothesis is not particularly convincing. Presumably the enormous temperatures at the earth's core must have reduced all its matter to a uniform liquid state. That is, it is unlikely that any solid masses remain in the liquid core to produce this shifting. Scientists have also considered the possibility that influences from outer space may be responsible for the fluctuations.

One theory states that the mutual gravitational attraction between various masses (the so-called "gravitational constant") is related to the state of all other masses in space. A disturbance of the gravitational constant might briefly disrupt the earth's rotation. This notion is pure speculation. Presumably, abrupt changes in the gravitational constant would have some visible effect on other planets besides the earth; but no such effects have been observed. Of course, these effects might be too slight to be detected by our instruments. Nowadays we are inclined to feel that no problem is too difficult for scientists to solve.

Yet no one has yet discovered the true cause of this relatively minor "mechanical" disruption of the earth's rotation. How little we really know about the world—even about our immediate surroundings!

These brief irregular accelerations and decelerations seem to have little effect on the earth. Nevertheless, they may indirectly influence all our lives. Some scientists suspect that, due to inertia in the earth's crust, the fluctuations in speed create areas of tension which may increase the danger of earthquakes on the earth's surface. It would be surprising if such a relationship did not exist. In order to test this hypothesis, scientists would have to discover whether earthquakes become more frequent directly after the fluctuations occur. We have already noted that more than 100,000 earthquakes of varying sizes are recorded each year. Only a complete statistical analysis could prove that some relationship exists between the fluctuations and the frequency of earthquakes. Here the time factor crops up again; not enough time has elapsed yet for us to establish conclusive statistics.

There is a second exception to our general rule that short-term fluctuations can be more easily diagnosed than long-term ones. The major perturbation of the earth's rotation is the so-called "secular" deceleration. Presumably this constant long-term deceleration will continue far into the future. Despite the fact that it is such a long-term phenomenon, scientists have been able to fully explain it. The factor which is slowing down the earth and will one day bring it to a complete halt—is the moon! The lunar mechanism which acts as a brake on the earth is known as "tidal friction."

No heavenly body travels alone through space; no heavenly body possesses a truly independent orbit. The sun's mass greatly exceeds the mass of all the rest of our solar system put together. Therefore the sun's gravitational attraction controls all the planets. But in turn the sun's orbit is controlled by a still larger mass: a region between the constellations of Scorpio and Sagittarius, which apparently forms the center of our galaxy.

The sun holds the earth and the other planets on course; in the same way the earth's attraction maintains the orbit of the moon. The moon is far smaller than the earth. All the same, it exerts some gravitational pull on our planet. When the moon is hovering over a particular point on the earth's surface, its gravitational force makes the earth lighter at this point. Our own weight and the weight of all other objects is determined by the force of gravity exerted by the earth's mass. Gravitational attraction draws everything toward the earth's center of gravity, located at the center of the globe. But when the moon soars high above our heads, its gravitation is pulling us in the *opposite* direction. On the surface of the moon, the lunar gravitational pull amounts to only one-sixth that of the earth. We have all observed the "unearthly" movements of the astronauts on the television screen; so we know how weak lunar gravitation appears compared with that of the earth. Moreover, the strength of the moon's gravitational pull decreases with "the square of the distance." That is, as the distance of a body from the point of origin of the gravitational force is doubled, the force is reduced to one-quarter of its strength (rather than to one-half, as one might expect).

Thus the moon's gravitational attraction is extremely weak by the time it reaches the earth. The moon reduces the weight of our bodies and of the objects around us; but the weight reduction remains so slight that we never notice it. All the same, the moon's gravitation is powerful enough to effect major changes in the surface of the earth. For one thing, the moon creates the earth's two oceanic tides.

The diagram on the following page depicts these tides. (The continents have been omitted in order to simplify the diagram.) We see that the moon actually produces *two* tides—a fact which confuses many people. One of these tides rises toward the moon; we have no problem seeing why this should be so. Another tide

1 (OVERLEAF)
The earth as viewed from the moon. Our planet has a diameter of only 7,900 miles, yet the entire history of life as we know it has unfolded on the tiny surface of this sphere. Only the thin outer layer of the earth, known as the "ecosphere," can support life. On the scale of this photograph, the ecosphere has a height of only about .001 inch. On this same scale, Mt. Everest would be barely .003 inch high; a person might just be able to feel it with his fingertip.

2
The Orion Nebula, a cloud of interstellar matter within our own spiral galaxy. The cloud lies 1,600 light-years from the earth and is 50 light-years across at its widest point. The Orion Nebula has a density one million times *thinner* than that of any artificial vacuum we can manufacture on earth. The cloud consists almost exclusively of hydrogen gas. Radiation emitted by neighboring stars ionizes the hydrogen, causing it to glow.

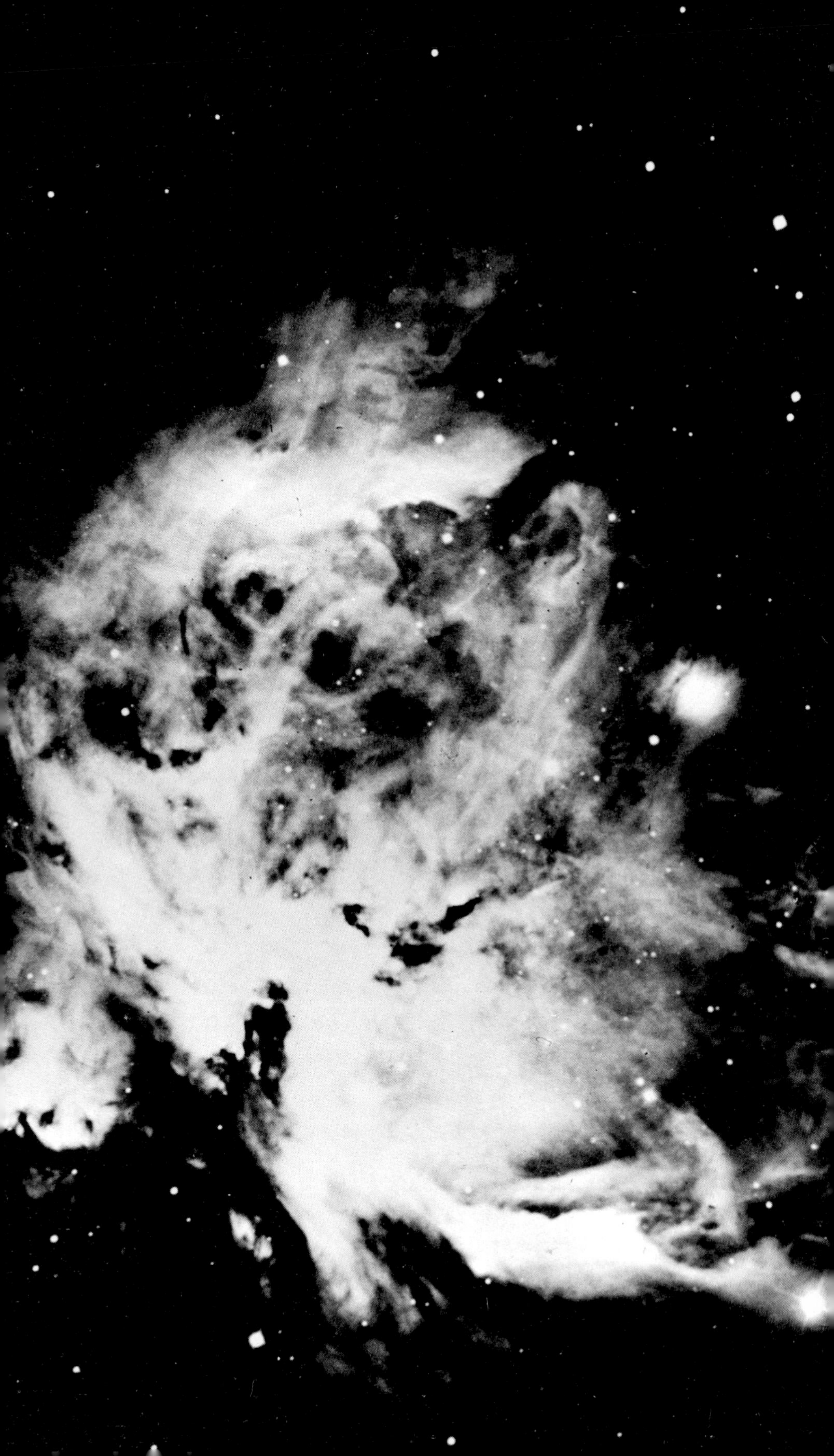

3

The famous Ring Nebula in the constellation Lyra. Like the Orion Nebula, this gaseous cloud lies within our own Milky Way Galaxy, "only" 5,000 light-years from the earth. This formation may actually represent the exploding cloud of a supernova. The small star in the center of the ring might have been the site of the original explosion. The speed at which the ring is traveling—it is still spreading outward at some thirteen to nineteen miles per second—appears to substantiate the theory that the cloud represents the remains of a supernova.

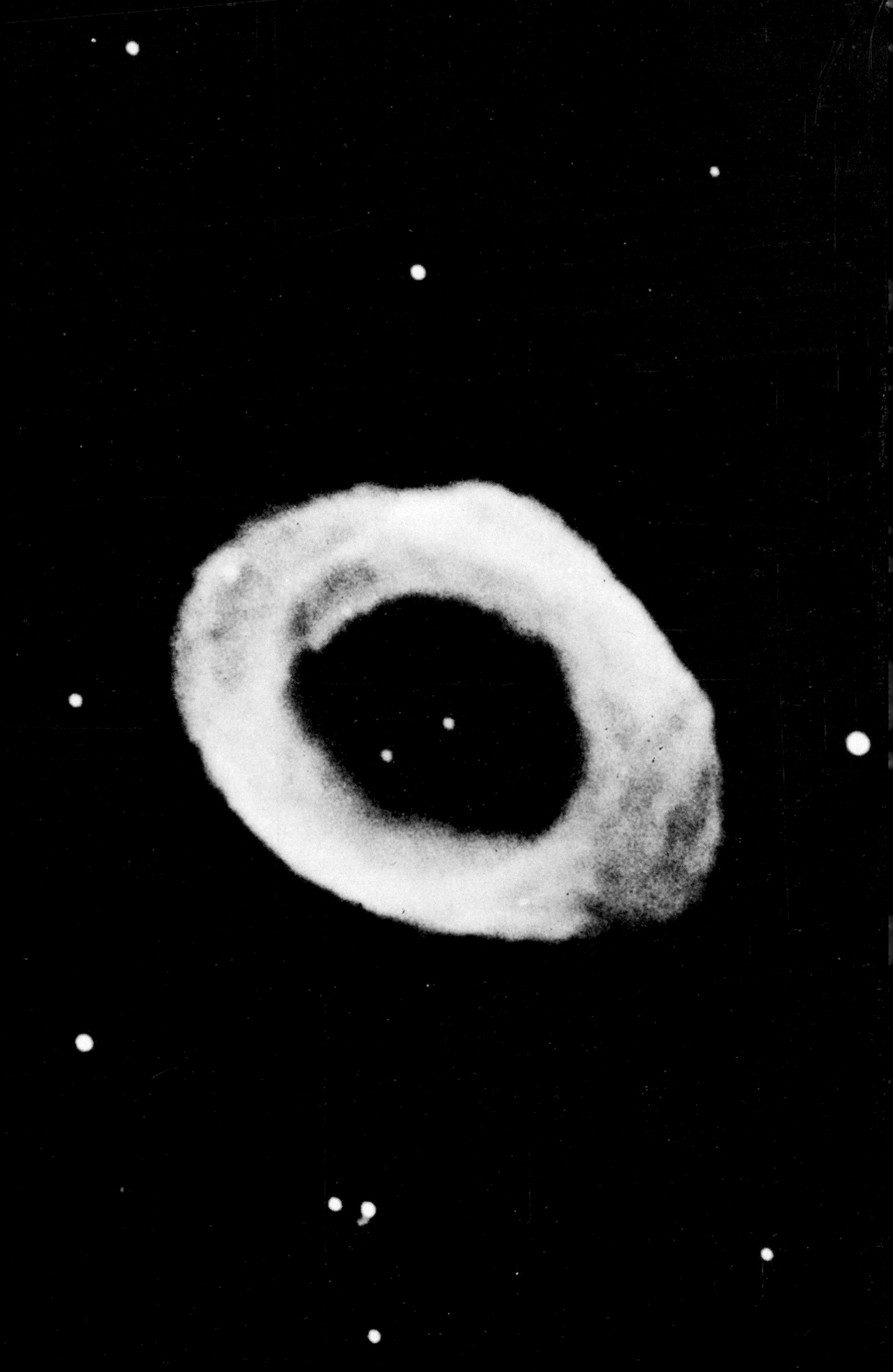

4

Typical spiral galaxy, an independent stellar system lying outside our own Milky Way Galaxy. This spiral system is located in the constellation of Canes Venatici, around 9,000,000 light-years away; thus it is one of our nearer galactic "neighbors." It is situated at such an angle in space that we are able to look directly down upon it, so that we have a clear view of the entire diameter of the galaxy. This stellar system contains at least 50 billion stars. If we could view our own galaxy from the same distance and from the same vantage point, it would look very much like this picture. Our own sun would be located about two-thirds of the way from the center to the edge of the galaxy.

5

The renowned Andromeda Galaxy, another of our "neighboring" galaxies. This neighbor consists of some 50 billion stars and lies some 3,000,000 light-years away. Viewed from the earth, the Andromeda Galaxy appears to be tilted in space; that is, we observe its surface from an oblique angle. The two blurred pea-sized spots represent "dwarf" milky ways bordering upon the Andromeda Galaxy.

6

Side view of a spiral galaxy. Like the Andromeda Galaxy, it consists of some 50 billion stars. It is about 100,000 light-years in length. Its width at the center amounts to some 12,000 light-years, but the thinner outlying disk area is only 3,000 light-years wide. Dark masses of cosmic dust have accumulated in the equatorial zone.

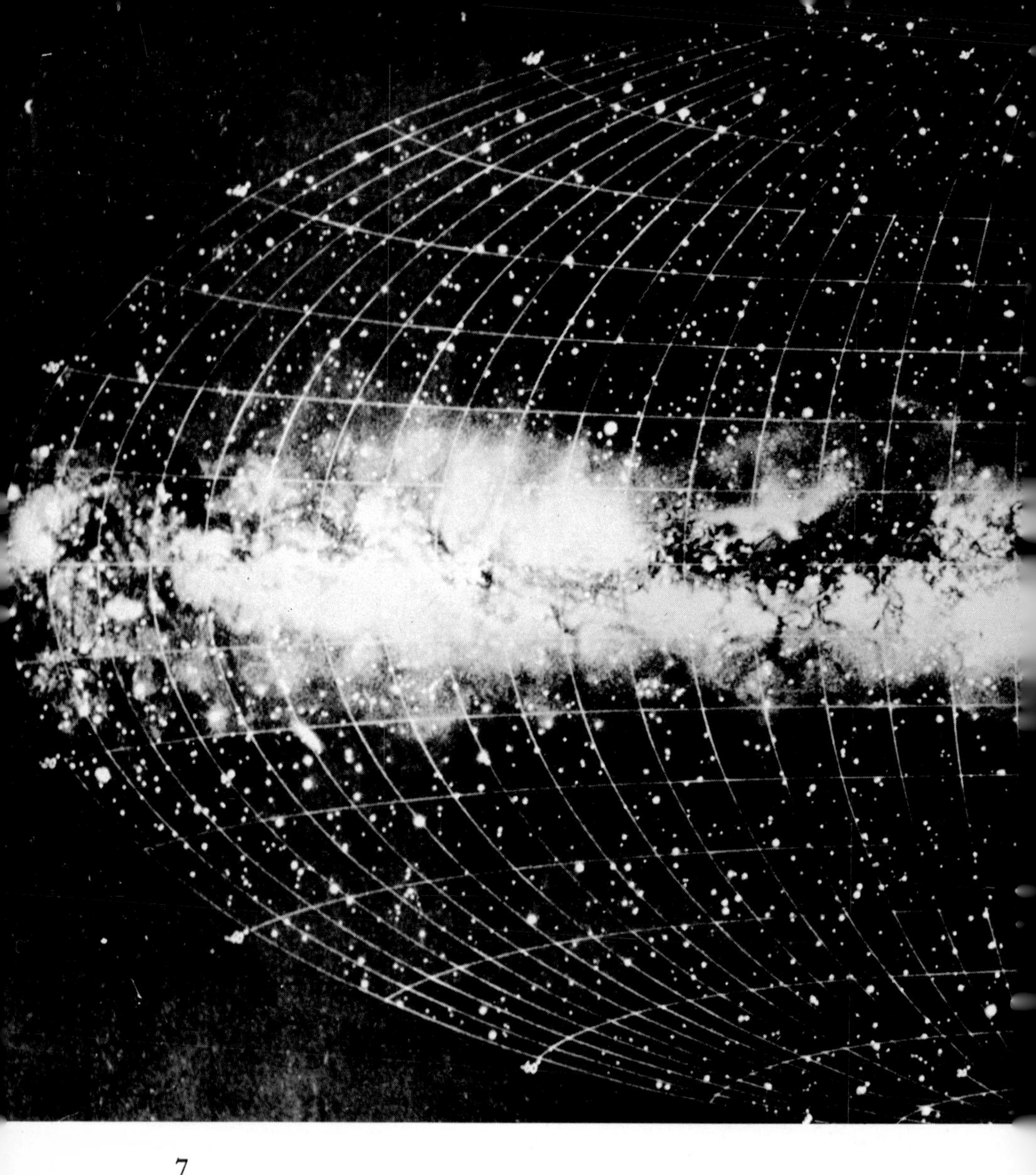

7
Panoramic view of our Milky Way Galaxy. Kant reached the brilliant conclusion that viewed "from outside," our Milky Way must be shaped like a giant lens. The Königsberg philosopher also correctly guessed that our galaxy was exactly like all the other spiral galaxies which had been observed in the heavens. He reasoned that we were on the inside of a galaxy looking out, so that we saw this galaxy only as a band of stars

encircling our earth. In modern times it has been demonstrated that our Milky Way Galaxy does in fact represent a giant spiral galaxy composed of at least 100 billion stars. Among these stars is our own sun. Moreover, astronomers now know that the characteristic "band" of stars around the earth (the Milky Way) is just what Kant supposed it to be: a spiral-shaped galaxy viewed "from the inside."

8

The Crab Nebula in the constellation of Taurus. This cloud represents the remains of an exploding supernova which destroyed an entire star. In the year 1054, Chinese astronomers observed the sudden blaze produced by this explosion. The Crab Nebula lies some 4,000 light-years from the earth. Nine hundred years after the explosion of the star itself, the exploding cloud is still expanding in all directions at a speed of more than 620 miles per second. The eruption of a supernova heralds the death of an individual star, but at the same time initiates the birth of a new generation of stars which exhibit properties different from those of their parents. Astronomers have learned that our solar system belongs to a relatively youthful star generation. Moreover, this succession of stellar generations is a basic precondition for life itself; if these various generations had not existed, our solar system, our earth, and we ourselves would never have come into being.

9

Nova Persei in 1901. The light surrounding the exploded star shines rather dimly. Apart from this fact, Nova Persei at first looks very much like any other nova. This nova lies 3,000 light-years from the earth. From this distance, it is difficult to tell that the cloud is still rapidly expanding. When astronomers measured its rate of expansion, they were astonished to find that the cloud appeared to be spreading in all directions at the speed of light itself! According to the theory of relativity, matter cannot attain the speed of light. An expanding gas cloud is composed of matter. Therefore the cloud of Nova Persei appeared to be contradicting the laws of physics. Eventually astronomers solved the mystery: What appeared to be an expanding cloud was not actually matter, but the light from the explosion itself! The text describes how scientists arrived at this conclusion.

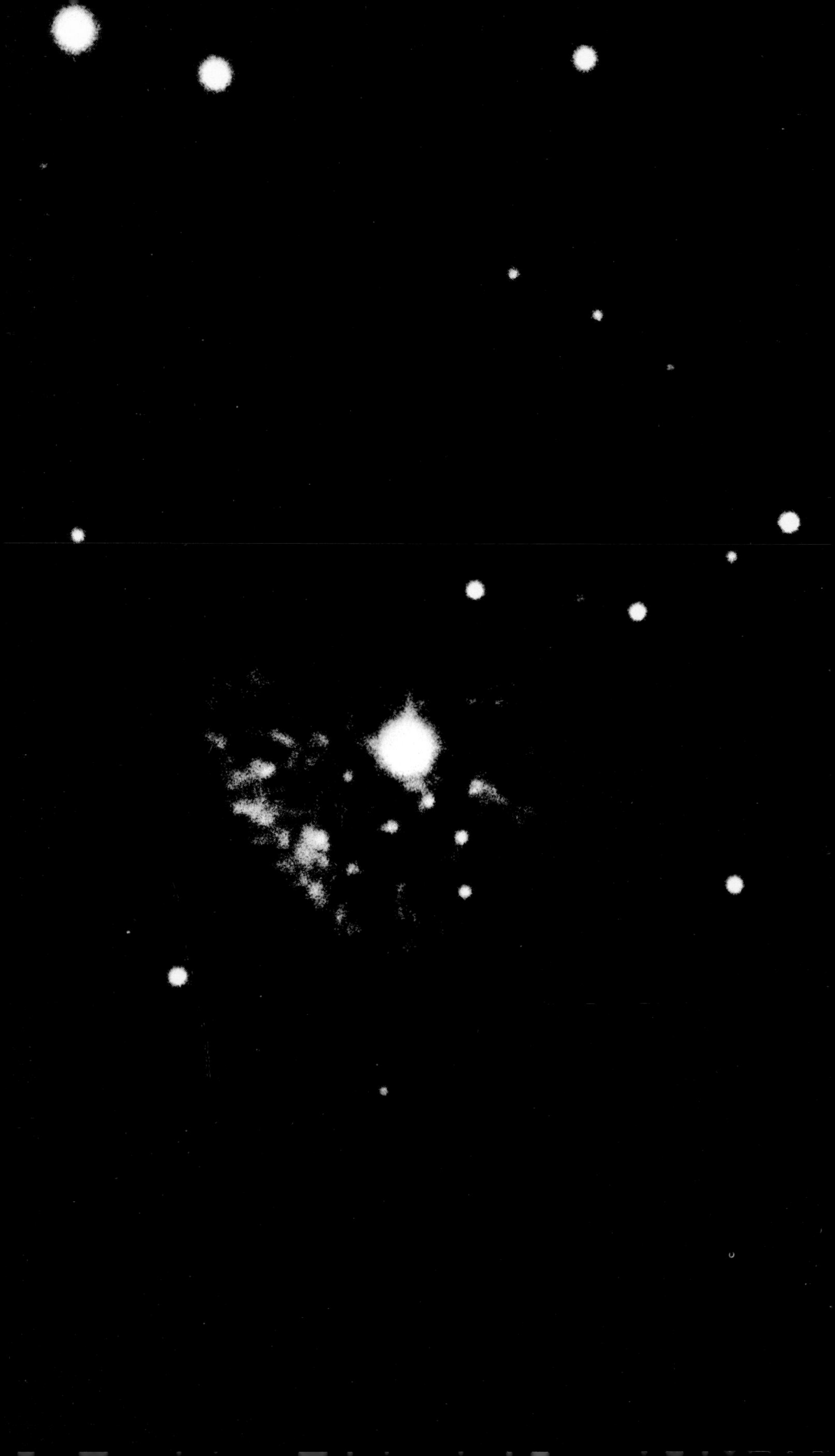

10

As the quality of telescopes continues to improve, astronomers discover more and more galaxies outside our own Milky Way Galaxy. This photograph shows four such galaxies, located some 20,000,000 light-years from the earth. The photographer could not avoid filming many "foreground stars" along with the galaxies. Foreground stars belong to our own galaxy; but since all photographs are taken from within our own Milky Way, some foreground stars always find their way into the picture.

11

This photograph contains more galaxies than foreground stars. The galaxies can be recognized by their elliptical shape and blurred outlines. All the galaxies in this picture are about 300,000,000 light-years from our planet. Each of them contains some 20 billion, 100 billion, or 200 billion suns and has a diameter of from 50,000 to 100,000 light-years.

12

Only a few foreground stars are visible. Almost every one of these teeming points of light represents an entire world—a galaxy the size of our own. It takes light almost one billion years to travel from the smallest of these dots to us on earth. Moreover, each of these far-off galaxies probably swarms with countless unimaginable forms of life.

13

Our sun with its characteristic sunspots. Kant and Herschel both had theories about these dark-looking spots. Kant believed the spots to be mountain tops projecting through the sea of liquid fire on the sun's surface. In reality these spots are not dark at all, but are hotter and more brilliant than white-hot steel. If we could remove one of the spots from the sun and place it in the sky, it would light up the earth as brightly as a full moon. Yet sunspots always appear black in photographs or when observed through protective filters. The text explains the reason for this misleading phenomenon.

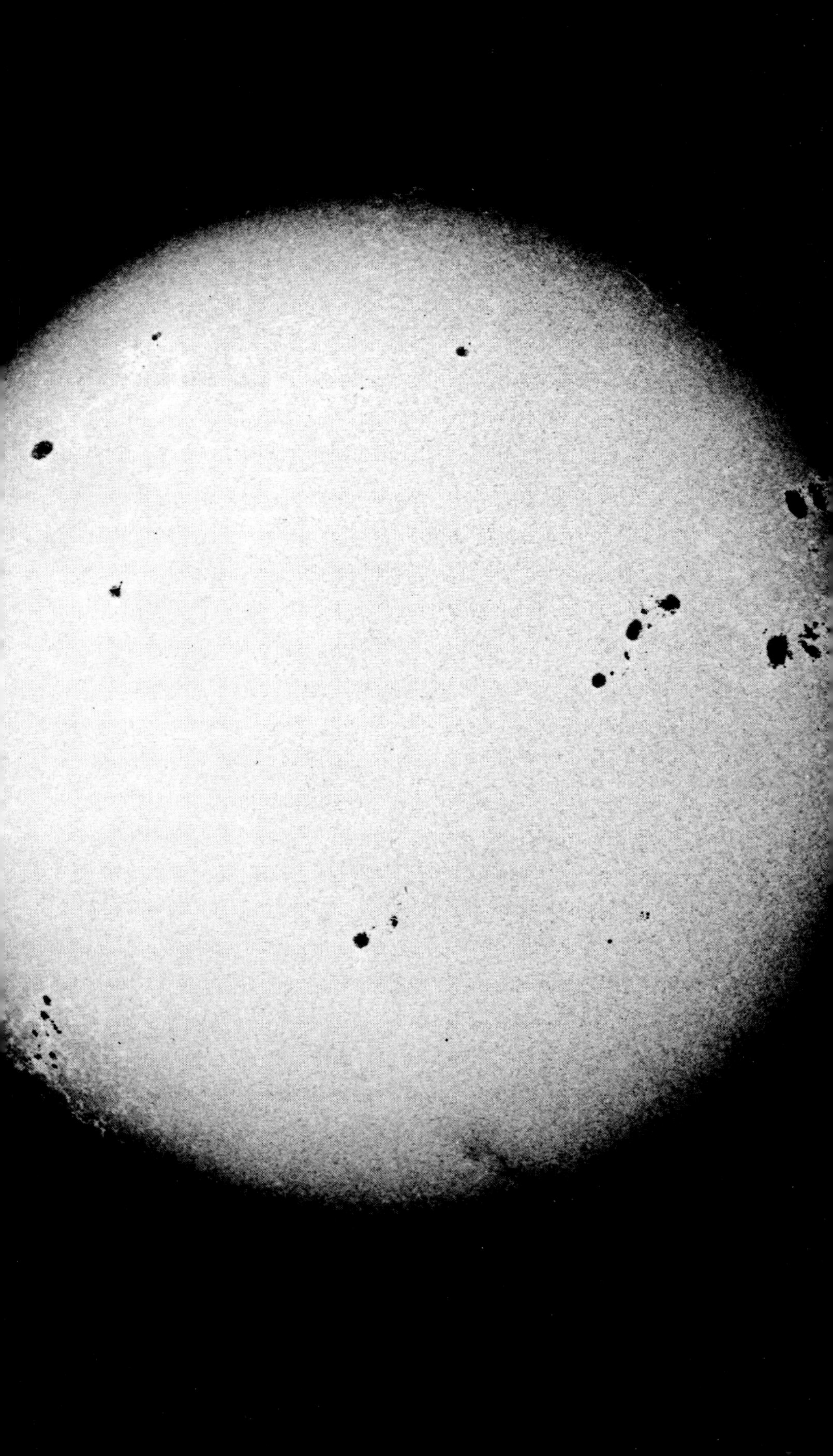

14

An unusually large group of sunspots, highly magnified. These spots are almost 32,000 miles across at their widest point, four times the diameter of the earth! The apparently black regions at the center of each spot are actually burning at temperatures of over 4,000° Celsius. The shaded zone around the black center is known as the "penumbra"; its temperature is about 5,300°C. The "granulated" surface of the sun itself burns at 5,500°C. Astronomers still know very little about sunspots. They have learned that the spots are associated with strong magnetic fields. The matter composing the sun's surface flows out from the spots at speeds of up to several miles per second. Astronomers have not yet discovered why the frequency of sunspots seems to follow an eleven-year rhythm, but this rhythm clearly causes fluctuations in weather conditions on earth.

In the last decade astronomers have learned that the sun emits another form of radiation besides light and heat, the so-called "corpuscular radiation." This radiation consists of atomic nuclei and electrons hurled from the sun at great speed. Scientists now believe that it decisively influenced the course of evolution on earth.

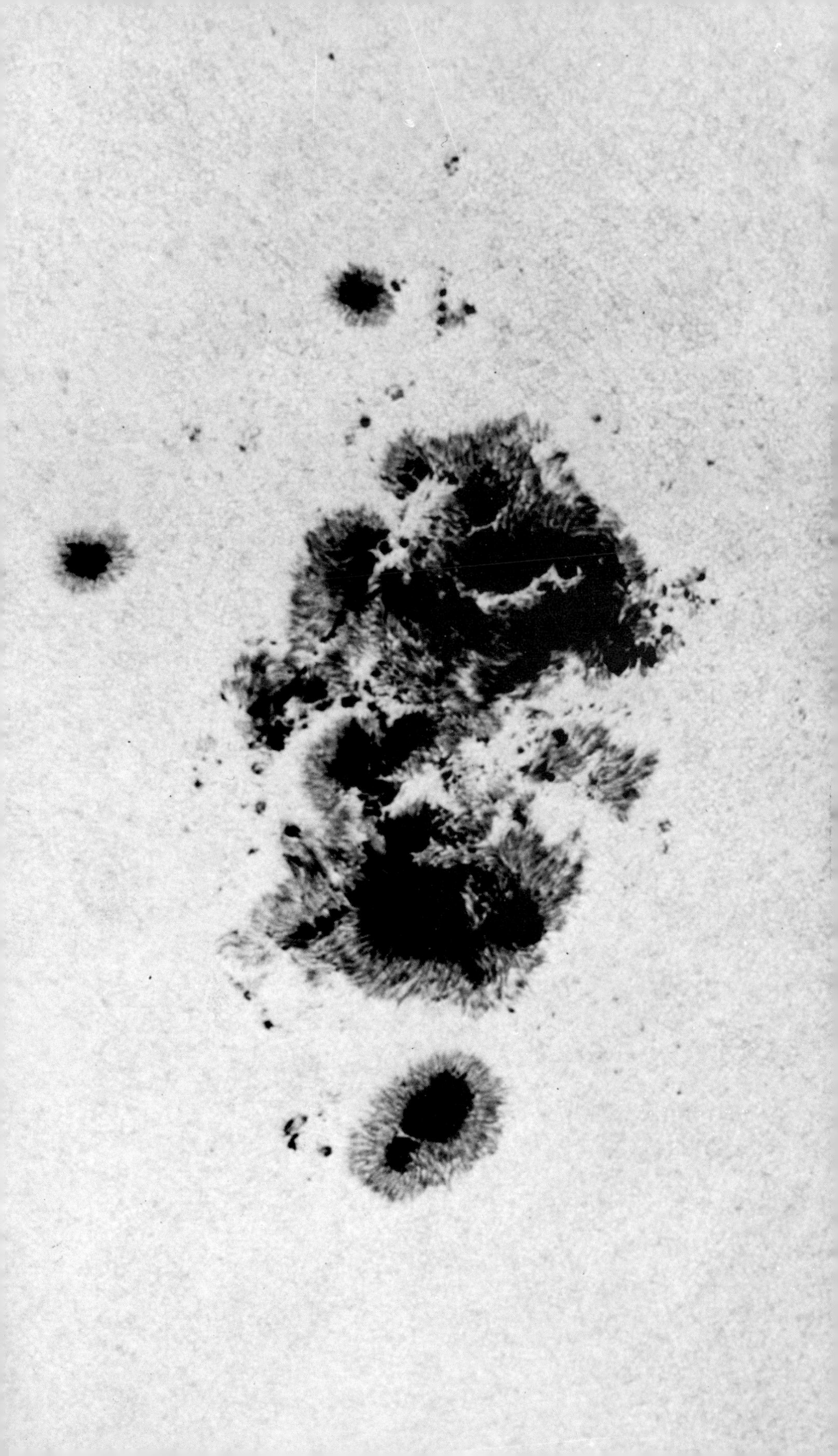

15

Petrified remains of an ammonite, a squidlike animal which lived 500,000,000 years ago. A few decades ago, fossils of this sort were a source of considerable embarrassment to astronomers. When the ammonite fossils were found, astronomers had not yet discovered that a star maintains its energy by means of nuclear

fusion reactions. They believed that our sun had only existed for a short time; if it had been very old, it would already have burned itself out. But various fossils contradicted this theory. The ammonites prove that our sun has been shining with undiminished intensity for at least 500,000,000 years.

16

Morehouse's comet of 1908. Its tail was about 19,000,000 miles long. Down through the centuries, astronomers began to notice that comet tails always point *away* from the sun. This was their first clue to the fact that the sun was emitting some mysterious force which drove the comet tails away from its surface. A few years ago scientists finally discovered the nature of this force.

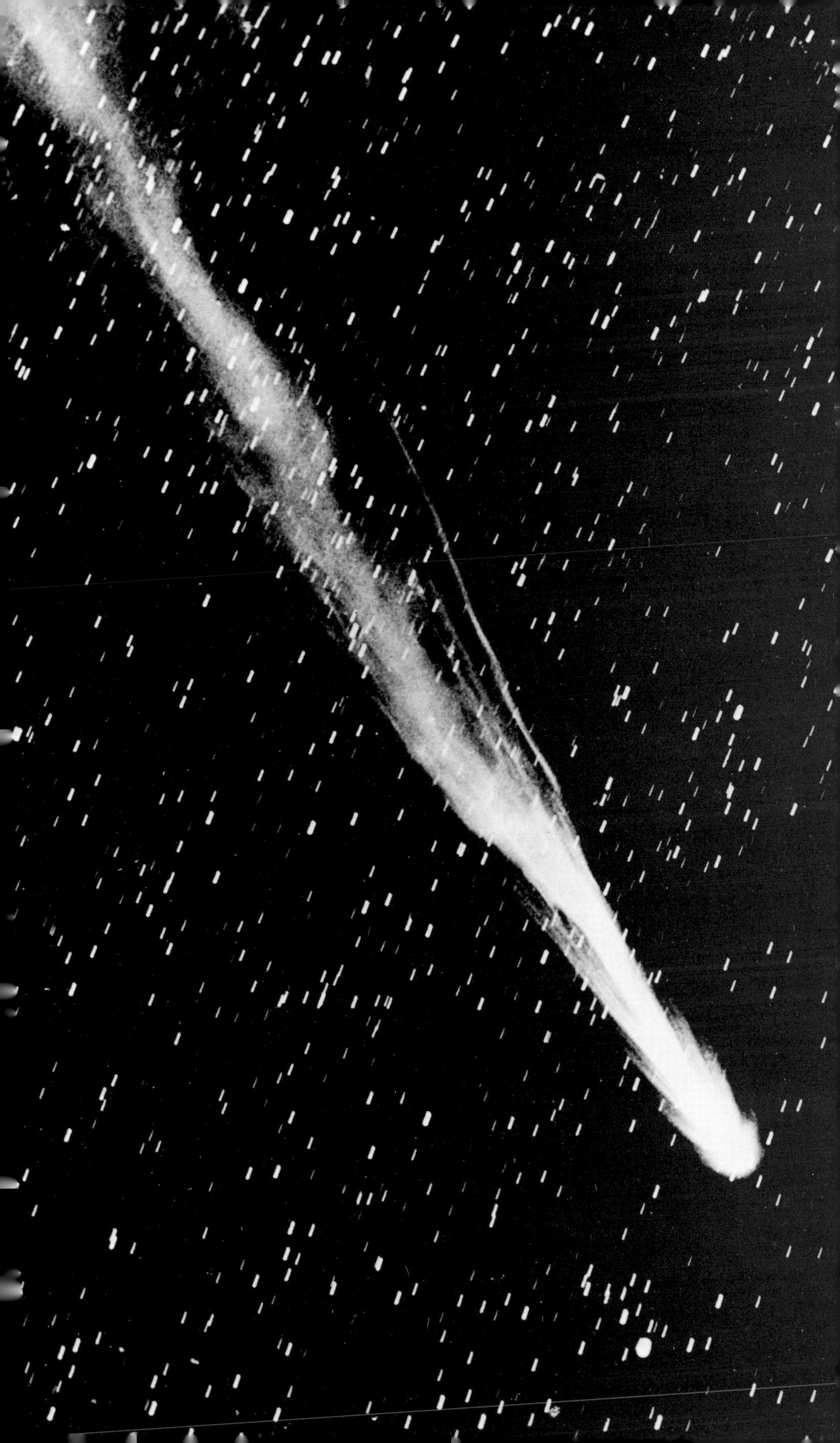

17

Enlarged photograph of a particularly powerful eruption on the sun's surface. The edge of the sun projects into the picture at the right. (To make this photograph possible, the sun itself has been blacked out.) Glowing streaks of gas are shooting from the sun and curving down toward the left. These streaks of gas are being hurled some 212,000 miles from the sun's surface—almost the distance separating the earth from the moon! On the scale of this photograph, our earth would be less than .2 inch wide.

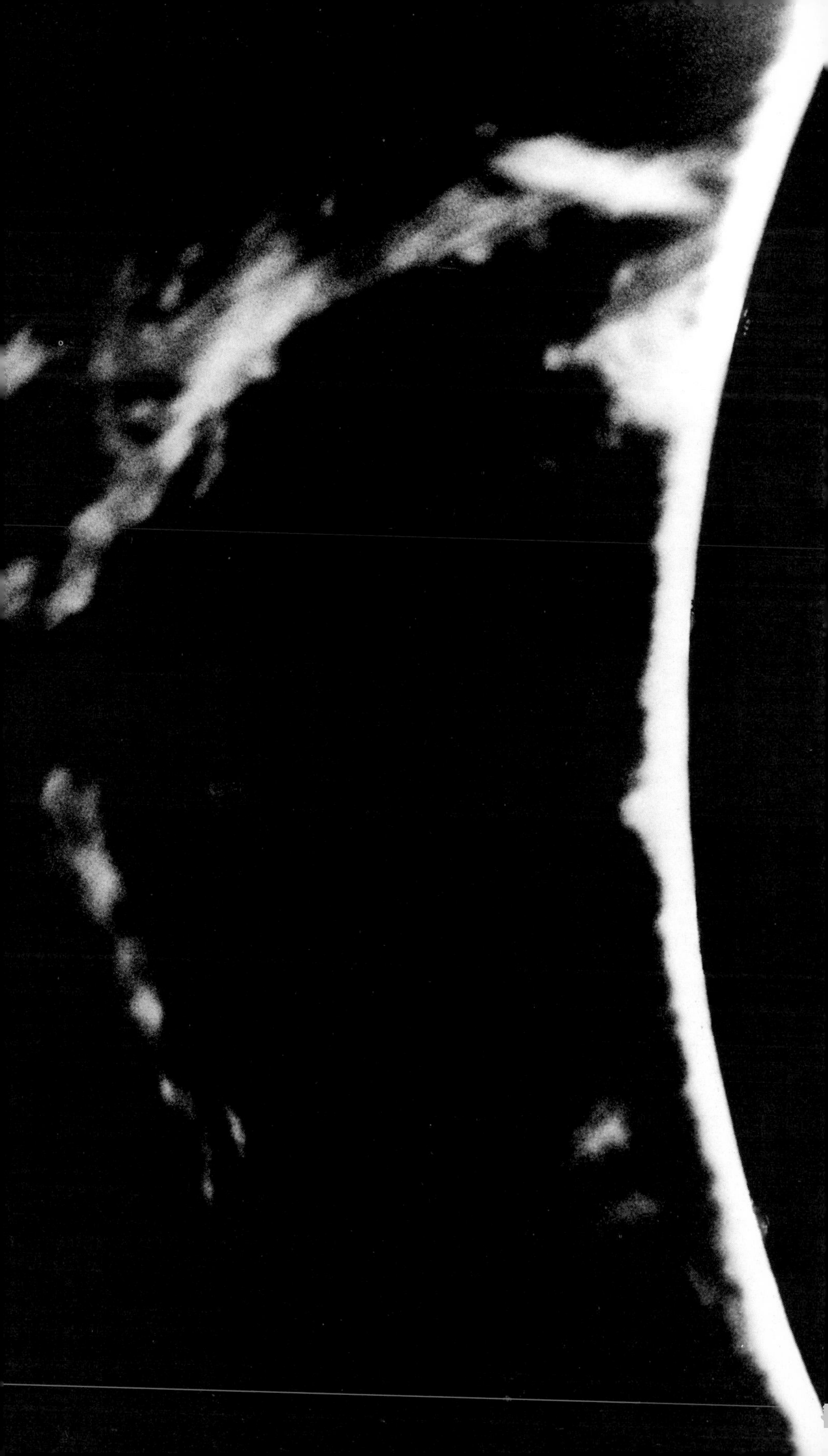

18

During a total solar eclipse, the moon completely covers the disk of the sun, so that for one brief moment the sun's corona becomes visible behind the moon. This corona forms the visible part of the sun's atmosphere; it rages at a temperature of 1,000,000°C. Recent radio-astronomical investigations and space probes have revealed that the sun's atmosphere extends far beyond the visible corona—*at least* past the orbit of Mars!

Until recently scientists were unaware that the sun shelters all the inner planets inside its own thin atmosphere. Moreover, there is every indication that the sun's atmosphere actually envelops our entire solar system. If the earth were not sheltered in this protective atmosphere, there would be no life on our planet.

19
Highly magnified image of the sun's surface, showing its characteristic "granulated" texture. The outer layer of the sun displays a sort of "honeycombed" structure, each "cell" of the honeycomb being over 620 miles wide! Scientists now believe that the "cells" represent gas bubbles rising from the sun's interior, producing a deafening noise. A constant thundering is produced by the bursting bubbles as they rise from the depths of the sun. Scientists believe that the sound waves heat the sun's atmosphere to 1,000,000°C, thus creating the energy to hurl the particles of the solar atmosphere out into space. Thus the sound of the bubbles seems to be responsible for the fact that the sun's atmosphere extends to the limits of our solar system. Indirectly, we owe our lives to this sound.

20a
Each day a coral deposits new layers of calcium at its rear (dark cone-shaped area of the diagram). The coral sits on this calcareous deposit as if it were on a sort of socket or pedestal. The animal does not produce the calcareous deposit at a uniform rate, but according to a seasonal rhythm.

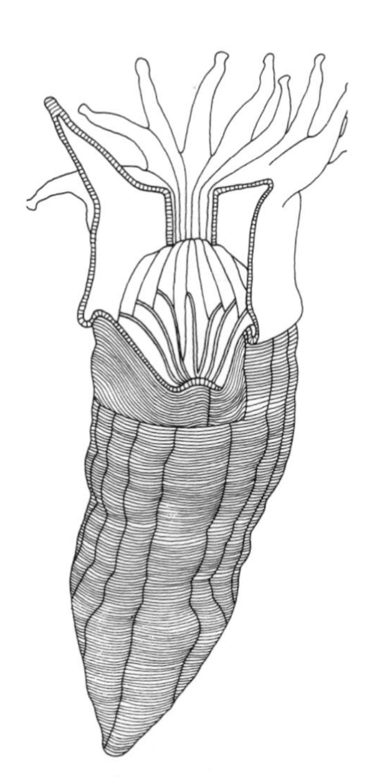

20b
Coral shell from the Devonian period. This photograph clearly reveals the various layers deposited according to a seasonal rhythm.

20c
Magnified image of the fossilized coral layers. This photograph shows the deeply cut annual rings. Still finer lines can be detected within each annual ring: These are the day rings, visible only under a microscope. Scientists counted the day rings and made an astounding discovery. Three hundred and seventy million years ago, when the corals formed these rings, each year must have had 395 days instead of 365! Fossilized corals prove that the earth's rotation has been gradually slowing down. The moon is responsible for this slackening of speed. (See the text for details.) The speed of the earth's rotation profoundly influences our surroundings and our own capacity for survival.

21a

Radiolarian skeletons magnified 500 times. Recently scientists have investigated the fossilized skeletons of these one-celled microorganisms. Each radiolarian species and subspecies manufactures its own type of skeleton. The study of these skeletons has proved that the dipolar reversals of the earth's magnetic field played a major role in the evolution of life.

21b

Part of a radiolarian skeleton, magnified 2,000 times. The photograph shows details of the surface structure. Scientists use these structural details to classify each individual radiolarian species.

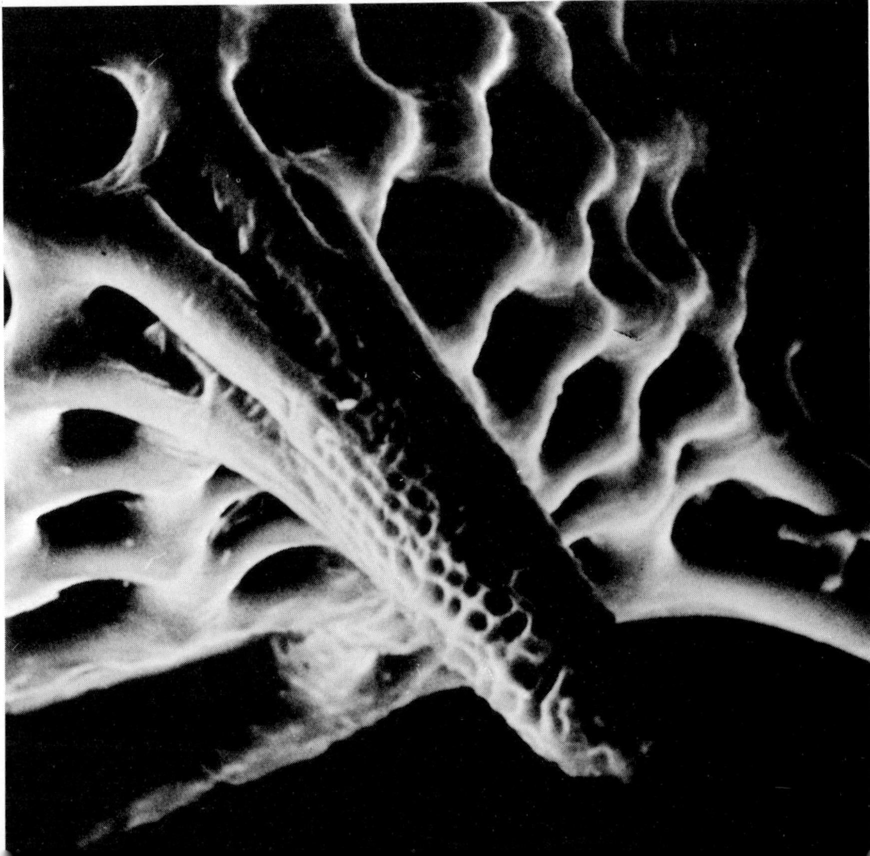

22
Typical tektites from various regions of the earth. These stones were formed from a molten vitreous mass; they have an average diameter of from one to two inches. Large numbers of tektites have been found in four specific regions of the earth. None of these four regions showed signs of earlier volcanic activity. Therefore these stones continued for years to puzzle the world's geologists. Recently a long chain of evidence revealed the true source of the "molten drops": They were produced by long-ago collisions of the earth with giant meteors. This discovery proved especially significant for a reason unrelated to the tektites themselves; the collisions between meteors and the earth strongly affected the evolution of life on our planet.

23
The Nördlinger Ries, a flat craterlike basin located in Germany between the towns of Dinkelsbühl and Do-

nauwörth. This depression was formed 15,000,000 years ago when a giant meteor crashed into the earth.

24
Scale model of the Nördlinger Ries, accentuating the surface relief 3.7 times. This model reveals the similarity between a meteor crater on earth and the craters of the moon.

25

These five slightly blurred patches of light are actually extremely remote galaxies located from 200,000,000 to 300,000,000 light-years from the earth. Glowing clouds of gas lying between the various galaxies have established a remarkable relationship between the five stellar systems. Apparently we are looking at an actual example of the intergalactic exchange of matter—the "metabolism" of the cosmos! Recently astronomers have discovered other examples of this phenomenon. If the exchange of matter did not take place on a truly cosmic scale, life would probably never have developed anywhere in the universe. The text explains the phenomenon of exchange of matter in greater detail.

26

Earlier photographs showed suns destroyed in supernova explosions. But at times this fate may befall an entire milky way! In 1962 astronomers detected the extremely remote galaxy shown here. Apparently the density of stars in the center of the galaxy exceeded

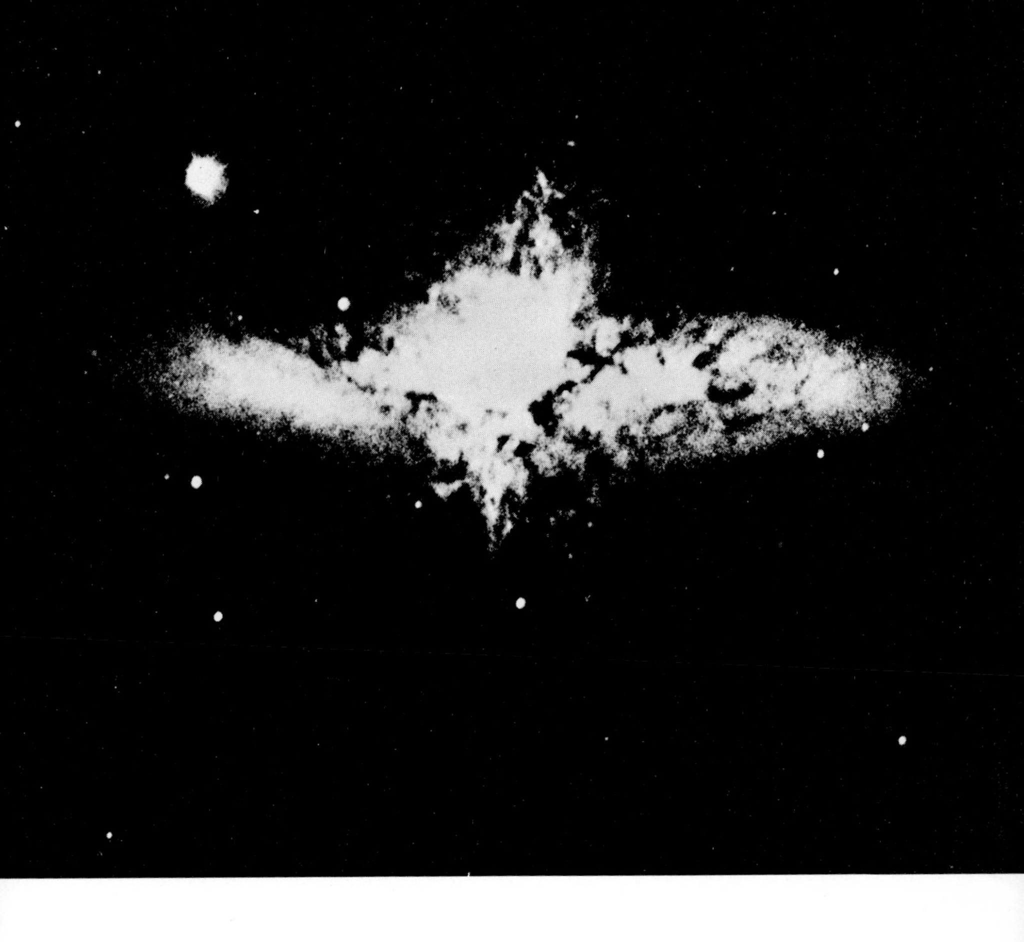

the critical mass, triggering a chain reaction of exploding supernovas. This chain reaction has been in progress for at least one million years. By now the entire central region of the galaxy has been blown to smithereens.

27
Typical spiral galaxy with the characteristic spiral arms. About sixty percent of the galaxies so far investigated fall into this class, among them our own Milky Way Galaxy. The spiral structure of spiral galaxies appears to be a basic prerequisite for the development of life as we know it.

28
A typical elliptical galaxy. Astronomers believe that galaxies of this shape have never produced life. Elliptical galaxies contain the same number of stars as all spiral galaxies, including our own, but probably no living creature will dwell there until the end of time. The chapter called "We Are Such Stuff as Stars Are Made Of" explains why a spiral-shaped galaxy seems essential to the development of life.

29

One of the at least 119 globular clusters found in our own galaxy. Each of these clusters is composed of hundreds of thousands of stars, distributed a thousand times more densely than stars in other parts of the galaxy. These globular clusters are the oldest formations in the known universe. They do not rotate with the rest of our galaxy, nor do they lie in the same flat plane as all the other stars in the galaxy. Instead, they are uniformly distributed on all sides of the galaxy, composing a giant sphere or globe. Astronomers believe that these stars were formed from an enormous spherical cloud which began condensing at least 10 billion years ago. Eventually this cloud developed into our present-day galaxy.

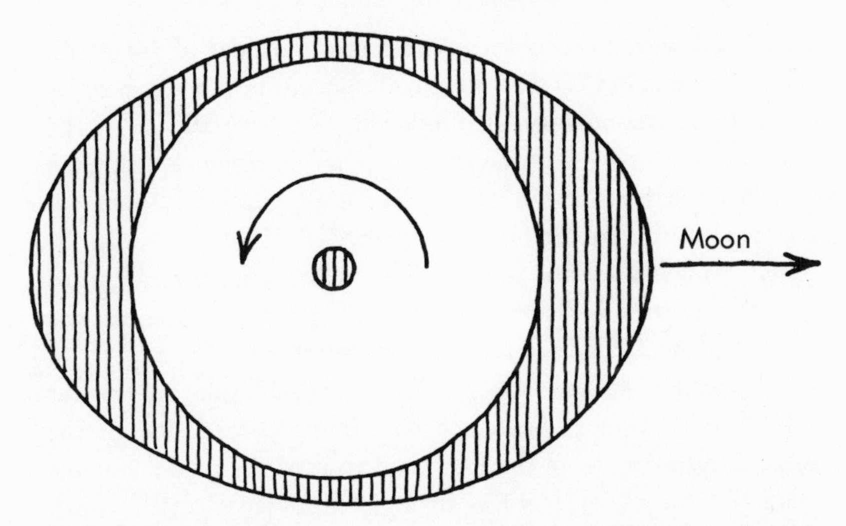

People always experience the earth as standing still. Therefore they always imagine that the ebb and flood tides flow around *the earth. In reality the water is held immobile by the moon's gravity while the earth keeps turning away beneath it. The earth expends force to pull away from the water; thus the tides continually slow down the earth's rotation. After a few billion years these tides will bring the earth's rotation to a halt.*

exists on the side of the earth that is turned away from the moon; but few people understand what creates it. We all know that high tide and low tide alternate in six-hour shifts. Clearly this is possible only if two tides exist. If there were only one tide, then any given point on the earth's surface could experience a high tide only once every twenty-four hours.

Thus a "paradoxical" tide exists on the opposite side of the earth. But what causes this phenomenon? A physicist can understand the whole matter quite easily; but it presents a problem to those of us who are not acquainted with the relevant mathematical formulas. We can simplify the explanation somewhat by referring again to our diagram. The moon does not attract only the water on the near side of the earth; it also attracts the earth itself and the water on its far side. But it exerts a far more powerful attraction on the near side of the earth than on the far side. Well over 7,000 miles separate the two sides of the earth. By the time it has crossed all this distance, the moon's gravitational force has been further reduced. We have already noted that the gravitational force decreases as the distance from the source increases ("the decrease of the gravitational force according to the square of the distance"). This point can be stated somewhat more simply. The water on the moonward side of the earth is *more* strongly attracted than the earth; the water on the far side is *less* strongly attracted than the earth. This phenomenon explains the "paradoxical" tide on the far side of our planet. The water on the moonward side of the earth flows away from the earth and toward the moon. But the water on the far side is less strongly attracted than the earth so that, instead of flowing toward the moon, it simply "lags behind" the earth.

We humans experience everything in terms of our daily lives. Thus we regard the earth as a stable and constant reference point for all our activities. The discoveries of Copernicus have done nothing to change this fact. We still think of the sun as "rising" and "setting"; we still speak as if the moon and stars moved from east to west across the sky—as if we were living in the age of Ptolemy! And yet we have known for centuries that the sun and the other stars are "fixed"; in reality the *earth* is turning from west to east. At bottom we believe only what we see with our own eyes. So we end up doing two sets of bookkeeping, one for our daily affairs, and one for purposes of "scientific" observation. In the second case we try to divorce ourselves from our narrow personal perspective.

Thus human beings always see two aspects of any situation simultaneously. This "double vision" is a heritage of our evolutionary history. The human brain enables us to perceive the objective reality of the world; but nature did not invent the brain to fulfill this function. The evolutionary process favors those biological mechanisms which aid a species to survive in its natural environment. Like all other organs, the brain developed because it contributed to the survival of a species. The human brain is an enigma. Originally it served a purely pragmatic function. Yet today we are capable of observing phenomena—the laws governing planetary orbits, or the tidal attraction between the earth and the moon—which bear no direct relation to our survival. How amazing that we should know anything of those realms which are unrelated to our survival as biological organisms! On the other hand, vast realms of nature (such as the world of subatomic particles) are virtually impenetrable to human beings. We can penetrate these realms only indirectly, using the crutches of abstract formulas and logical symbols. We may manipulate these symbols, but they remain empty concepts, alien to our senses. Thus it should not surprise us that, despite all we *know* about the stars, we still *experience* them as objects moving overhead.

Ultimately we always rely on the direct testimony of our senses. This explains why we picture the tides as water flowing around a motionless earth once every twenty-four hours. In reality both tides are held fast by the moon's attraction as the earth moves away beneath them. Not the water, but the earth must "flow." Most of us assume that the earth rotates in empty space, encountering no resistance. But actually, the earth expends energy in resisting the hold of the water. On the other hand, not all the water in the world is being held perfectly still by the moon. If this were really the case, then the earth would have stopped turning long ago. Or more precisely, the earth would rotate only once a month as the moon moved around it, drawing the earth along with the tides. The same side of the earth would always be turned toward the moon. Thus one earth day would last as long as a month does now; and the sun would rise and set only twelve times a year over any given point on the earth's surface.

Undoubtedly the moon will one day limit the earth's rotation to once a month. But this will not happen for another two or three billion years. The moon's braking effect on the earth takes place almost imperceptibly. The internal friction of water causes the oceans to move along with the earth as it turns. The water molecules composing the tides are continually being exchanged for other molecules. The same thing happens when a storm wave races across the surface of the sea. This wave does not actually pick up a given quantity of water and transport it across the sea. It simply causes all the water molecules it encounters to circulate in a rising and falling motion. The molecules hesitate slightly as they rise and fall. This hesitation causes a visible crest to form as the wave travels across the surface of the water. The visible crest does not represent an actual shifting of matter; rather the crest marks the temporary location of a constantly shifting vertical motion.

Thus the water on the earth's surface participates in our planet's rotation. If this were not true, the moon would impede the earth's motion to a far greater degree than it actually does. All the same, the moon does cramp our style as the earth performs its pirouette in space. The oceans *do* rotate along with the earth. But as the earth turns, its various continents keep crashing into the two tides held in the grip of the moon. When we stand on the coast, we see this phenomenon from the opposite point of view. It looks to us as if the tide were moving toward the land and breaking against it. Of course, what is important here is the amount of *force* produced by the encounter between sea and land, for this force is what slows the earth down. As far as the force of impact is concerned, it does not matter whether the land crashes into the sea or the sea into the land. A person may be in a car traveling at forty miles an hour when he runs into a parked car; or he may be sitting in a parked car which is rammed by another car traveling at forty miles an hour. Either way, the actual force of impact remains the same.

Now let us take another look at our diagram of the tides. We have noted that the earth expends effort in order to combat the friction produced by its contact with the tides. That is, the earth's initial impulse of rotation is slowly being exhausted. The ice skater's skates produce friction with the ice; they also encounter resistance from the air. Eventually the constant friction wears down the ice so that the skates no longer slide on it properly. The constant friction between the earth's surface and the tides produces the same effect.

The earth acquired its initial impulse of rotation some four or five billion years ago when our planet first came into being. Once any of this impulse is expended, it can never be regained. The earth rotates quite swiftly on its axis, as do most (but not all) of the planets. This swift rotation must be somehow related to the creation of the earth. We do not yet completely understand how this creation took place. We must assume that the earth arose out of a cloud of tiny particles which were widely distributed in space. Eventually the mutual attraction of the particles caused the cloud to steadily contract. That is, the earth probably developed in much the same way as the sun and other heavenly bodies.

Scientists still dispute the question of whether the earth was formed from a cloud of gas or a fine cloud of cosmic dust. The sun formed from a gaseous cloud but contains only very small amounts of such heavy metals as are found on earth. Scientists have not yet discovered the reason for this difference in composition between the earth and the sun. In any case, the heavy-metal content of the earth argues against its having formed from a cloud of gas. If the earth formed from cosmic dust, then the elements now composing this planet must have been present in the dust from the beginning.

Whatever the composition of the original cloud, it formed the nucleus of our earth. During our discussion of the sun, we observed the contraction of a similar cloud. We noted that such a contraction inevitably produces a carousel movement; the entire mass rotates. At first, the rotation is slow; it picks up speed as the cloud continues to reduce its diameter. As the contracting cloud of the earth shrank to a red-hot ball, it began to turn faster and faster. This acceleration ceased only when the glowing earth ceased to contract. Because of its relatively small mass, the earth never attained the temperature of the sun. Therefore it never developed the capacity to kindle atomic reactions in its core. Once the earth had ceased to contract, its rotation began to slow down again.

No force existed capable of increasing the speed of the earth's rotation once it had begun to slow down. The speed of a pirouette can increase only as long as the mass is contracting. When the spinning object has contracted as far as it can, the pirouette has attained its maximum speed. The cloud which formed the earth always had a limited energy supply. It merely exchanged large dimensions and slow rotation for small dimensions and fast rotation; its total energy remained the same. Once lost, not a fraction of the cloud's initial energy can ever be recovered. The effects of tidal friction may appear minimal, yet tidal friction represents a constant force which continues to play a decisive role in our planet's history.

It may surprise us to hear someone refer to tidal friction as "minimal." After all, that description of our continents crashing into the tides sounded rather violent. Actually the "crashing" does not amount to much, since the mass of the earth greatly outweighs the mass of the oceans composing the tides. The earth's mass exceeds that of the oceans by a ratio of about 4,000,000 to one. One of our earlier mental constructs can help us to picture this ratio more clearly. Let us imagine the earth as a polished billiard ball. We breathe on this ball so that our breath creates a thin film on its surface. On this scale model, the thickness of the film would greatly exceed the depth of the earth's oceans. In order to correspond exactly to the depth of the oceans, the film should be exactly .06 millimeter, or about .0023 inch.

A tide rising out of this more or less paper-thin layer of water can hardly have any radical effect on the earth's speed of rotation. But minimal as this effect may appear, it began at the moment the earth acquired a moon to control the tides. Scientists have discovered that each century the moon slows down the earth little more than a thousandth of a second per day (.00164 second per day, to be exact). Thus every one hundred years the days get a tiny bit longer. The change is imperceptible to human beings. But even this tiny change slowly accumulates. Those who measure time in terms of geological epochs can appreciate the role which tidal friction has played in the earth's history.

Two hundred million years ago the dinosaurs ruled the earth. At that time the years consisted of 385 days, not 365. The speed at which the earth orbits the sun remains constant; therefore the length of the year remains constant too. But when the earth rotates faster or slower on its axis, the days grow correspondingly shorter or longer. Although the length of the year remains constant, a greater or lesser number of days can fit into that year. In the age of the dinosaurs, the year contained 385 days, but each of these days lasted only twenty-three hours, not twenty-four. A year always consists of 8,760 hours. If these hours are divided among 365 days, then each day has twenty-four hours. But if the year has 385 days, then only twenty-three hours fall to each day.

The farther back we look into our planet's past, the faster the earth was turning. The younger the earth was, the less time tidal friction had been working to slow down its rotation. Consequently days in the past were shorter. About 400,000,000 years ago the first plants were leaving the water and beginning to grow in the coastal areas of the land. At that time a year must have had about 405 days, each of which lasted only 21½ of our hours. Some 600,000,000 years ago the earth was in the Cambrian Age; vertebrate animals had not yet developed. The seas were full of invertebrate life forms, some of which were just beginning to grow external skeletons. The days then lasted twenty hours, and 425 of them fit into a single year.

These figures represent more than mere theory. Scientists did not simply draw a few logical conclusions on the basis of present-day conditions. Everything we have mentioned was once a reality. The American scientist J. Wells recently made a startling discovery which directly proved that the earth's rotation has been slowing down. Wells devised a brilliant method of literally counting the number of days in a Devonian year. He gathered fossilized corals which various tests had shown to be about 370,000,000 years old. We know that living corals form their stone-hard armor according to a definite seasonal rhythm. Uniform rings form each year on the coral's surface, just as rings mark each year in the growth of a tree (see Illustrations 20a, b, and c). The American scientist examined his corals more closely and found what he had been seeking: Under powerful magnification the corals revealed definite "day rings" inside each of the yearly rings. These fine lines developed 370,000,000 years ago. Each time the temperature dropped and the light grew dim, the coral stopped producing its calcareous deposit for the night; another day ring formed in the coral's armor. When Wells counted the day rings, he found exactly 395 of them in every single one of the annual rings. He had in effect discovered the number of days in each year when the corals were still living creatures—before tidal friction had taken its toll of the days.

The Biological Clock

WE HAVE seen that one day the lunar tides will completely harness the earth's rotation. When this happens, the same side of the earth will always be turned toward the moon. A single day will last as long as one of our months, so that a year will have only twelve days. Each of these days will consist of two weeks of sunshine followed by two weeks of total darkness. The length of time the sun shines will vary somewhat according to the season, just as it does today. These seasonal variations will prove fatal to all living creatures.

It may seem idle to speculate over the fate of the human race in this advanced stage of earth history. Indications are that our race will have died out long before then. Natural history teaches us that not only individuals but whole species pass through the stages of youth, maturity, and old age. One-celled organisms such as bacteria, algae, and protozoa form the only exceptions to this rule. Yet despite the fact that our species will probably become extinct before the days get to be a month long, it may be useful to examine the effects which such extremely long days would have on human life. Examining these effects will help us to understand how profoundly we are influenced by the present speed of the earth's rotation.

Just think for a moment how refreshed we all feel by the evening breezes which follow a hot summer day. Our summer sun shines for sixteen hours a day at most. Just imagine what two weeks of unremitting sunshine would be like! Temperatures would steadily rise on the sunny side of the earth. Human beings could not survive these temperatures unless they were housed in special climate-controlled enclosures and wore protective clothing. But while they were safe inside, the animals and plants would be decimated; countless species would become extinct. Such conditions would completely upset the balance of nature! During the winter, the dark side of the earth would suffer just as severely. As two weeks passed without a single ray of sunshine, arctic conditions would ensue.

While the earth was undergoing these violent fluctuations, the moon too would have its day and night. During the lunar day, temperatures now rise to some 120° Celsius; the nights are −130°C. Of course, even when the earth's rotation is completely dominated by the moon, the earth will not undergo quite such extreme temperature variations as the moon does. The earth's atmosphere will somewhat diminish the intensity of the sun's rays, whereas the moon has no such protective atmosphere. Moreover, the earth's atmosphere stores up a certain amount of warmth during the day. This stored warmth is slowly dissipated during the night; it would help to take the chill off the arctic temperatures on the dark side of the earth. Thus the earth's atmosphere will help to soften the two extreme temperature ranges.

All the same, the situation will be pretty uncomfortable. Moreover, the "softening" influence of the earth's atmosphere will produce an unpleasant meteorological side effect. As the temperature changes, air will flow from the boiling-hot sunlit areas into the frigid nocturnal ones. Powerful air currents will move through the earth's atmosphere. To be sure, these air currents will help to equalize the temperature extremes. But they will also subject the earth to constant major hurricanes!

Long before the development of these extreme temperature changes, human beings would be affected by the atmospheric changes. Eventually earth days would begin to last for thirty or thirty-six hours. Very few visible changes would take place on the earth. Yet human beings would already be suffering acute discomfort. They would be suffering because of a phenomenon which scientists discovered only two decades ago: the so-called "biological clock." Although this "clock" regulates all our body functions, it is second nature to us. No one was aware of its existence until scientists began to study its effects.

Until recently everyone thought that the cycle of sleep and waking was controlled by the cycle of day and night. Even scientists had always assumed that animals and men automatically get tired and go to sleep when it gets dark. Then careful experiments were conducted; plants and animals were kept in rooms segregated from the outside atmosphere and lit with artificial lighting. The results showed that the twenty-four-hour rhythm of the astronomical day is innate to all living organisms, including plants.

Scientists raised these plants and animals in windowless laboratories. The subjects were kept in total darkness or in artificial light maintained at a constant level. The plants and animals had never been exposed to the outside world; their parents had been raised under the same laboratory conditions. Similar experiments were carried out with volunteer human subjects in underground bunkers. These human subjects remained sealed off for periods of many weeks. The experiments revealed that human beings, like animals and plants, maintain the twenty-four-hour cycle regardless of external conditions.

This innate cycle constitutes a sort of "internal clock" which greatly contributes to the survival of all biological organisms. This instinctive mechanism apparently helps living creatures to adapt to the various seasonal changes which occur in the course of each year. Scientists still have a great deal to learn in this new area of research. But they do know one thing: Nature has chosen an ingenious method of preparing her creatures to withstand seasonal alterations in their environment.

In recent years botanists have extensively investigated the role of the internal clock in the cyclical blossoming of plants. When we see the first flowers blooming in the spring, we may simply assume that the seasonal rise in temperature has triggered their sudden growth. Indeed, temperature *does* play a major role in plant growth. We all know from experience that flowers bloom later in an exceptionally cold spring than in a spring which arrives prematurely. But temperature is not the only factor which controls the blooming of flowers. On rare occasions the temperatures remain cold well into May or June, and it seems as if the summer will never come. Yet the spring flowers bloom anyhow, even if they are a bit late in getting started. Experiments with artificial lighting have revealed that the length of the days also affects plants. Even if the weather grows prematurely warm, the plants will maintain their guard until the daylight begins to last a certain number of hours each day. Thus plants have a sort of twofold "internal security system." Two factors—the temperature and the length of the days —unite to set off the flowering process. It would be difficult to imagine a more effective method of protecting plants against the danger of opening during a temporary bout of warm weather. To be sure, even this device is not foolproof; it cannot protect the plants against later fluctuations in the normal weather pattern.

This "time clock" device reveals a significant aspect of the life of plants. Experiments have shown that most plants are actually able to "measure" the length of the day with an accuracy approaching that of mechanical clocks. In fact, the plants vary by only a few minutes from our own clock time! All measurement implies comparison. In order to measure the length of the days, a plant must have some sort of internal standard by which to compare the days. Thus plants, like all other creatures, must possess an internal clock whose precision surpasses that of any clock built by human hands before the seventeenth century!

We do not know where this biological clock is located in various organisms; nor do we know what it is or how it really functions. Experiments indicate that the clock may reside in the cell nucleus. These experiments also suggest that the clock may be composed of certain basic chemical processes–so-called "enzyme reactions." The great regularity with which these processes take place might make them a suitable standard for the measurement of time.

Little as we know about this internal clock, we do know that it regulates the time mechanism we have observed in plants. The clock mechanism in each variety of plant is timed to react to a different period of daylight. Each type of plant will open at the moment most favorable to its own species. We all know that certain kinds of flowers typically bloom in the spring, others in the summer or fall. In 1922 archaeologists discovered the unopened tomb of the Pharaoh Tutankhamen in the Egyptian Valley of the Kings. When they opened the tomb, they found a bouquet of wilted flowers lying on the stone threshold. The flowers had almost turned to dust, but botanists were still able to identify them. Three thousand years had gone by since the Pharaoh's death, yet these flowers made it possible to determine the time of his burial. It must have been the end of March or the beginning of April!

In the world of nature everything has a purpose, even if human beings do not know what the purpose is. Therefore there must be a reason why different varieties of plants bloom in different seasons. Like all other living creatures, plants must compete for survival. They accordingly develop various means of "giving each other a wide berth." We all know that plants do not flower for esthetic reasons, but to insure their own survival. Because they cannot move about, plants must depend on a "go-between" in order to reproduce. Their most primitive solution to the problem was simply to scatter pollen so that it could be picked up by the wind. This method proved ineffective. The plant had no control over where the pollen went; most of it never reached a plant of the same species.

Eventually plants succeeded in "harnessing" flying insects to act as go-betweens. This method proved vastly more efficient than the former random dispersion of pollen. Insects fly from flower to flower because they have learned that this method provides them with an inexhaustible food supply. Their flight pattern is involuntary; they always fly directly from one flower to the next, carrying the pollen with them on their bodies. Thus they serve as ideal go-betweens. Virtually every single grain of pollen will find its way to some neighboring flower.

But this method was not quite efficient enough. Not one, but many varieties of plants exist. In order to achieve pollination, a plant must make sure that its pollen reaches other plants of its own kind. The pollen-bearing insects had no way of discriminating among various species of flowers. Thus large quantities of pollen were dispersed without result. Despite this disadvantage, the plants could probably have continued to survive. But for some mysterious reason, nature always labors to improve a species, even at the cost of increasing its complexity. (If this were not so, mammalian life would never have come into being; we ourselves would not exist. For the most part living creatures could function far more comfortably in the water than on land. If it were not for nature's tendency to vary the species life would probably never have left the sea. Even the dinosaurs were perfect in their way. But the evolutionary process did not stop with the dinosaurs either. Simple logic suggests that our species does not represent the last word in the evolution of life on earth. One day the evolutionary process will leave us behind and travel on to some new goal whose nature we cannot begin to imagine.)

Even after flying insects began to transport pollen, plants continued to develop more refined reproductive techniques. The more advanced plants now have methods to insure that most of the pollen picked up by visiting insects will be carried to plants of their own species. This fact explains why we now see such a diversity of plant types, all differing greatly in size, color, and shape. When we look at flowers, we first notice their beauty of color and form. In reality all this beauty represents an elaborate system of "recognition signals." Once plants had developed flowers, various insect species began to "specialize" in particular blossoms. The plants in turn produced increasingly striking and colorful blooms which "signaled" to passing insects. A flower might be far away and almost lost among plants of other species; yet the insect could recognize it at once by its characteristic bloom. Thus plants by now have discovered a more or less ideal solution to their problem. They have come near to achieving the controlled exchange of pollen among plants of the same species. No doubt plants are continuing to evolve. But they evolve so slowly that hundreds of millions of years must go by before any perceptible change can occur.

Nature devises various means of reaching the same goal. Certain insect species now specialize in just one or several species of plants. But in the competitive world of nature, a single trump card cannot guarantee the survival of a species. Therefore the various plant species have adopted the practice of blooming at different times. To be sure, a wide variety of flowers now bloom in the same season. Nevertheless, all the available time from the spring to the fall has been divided up as equally as possible among the various plant species. Each species looks for a kind of "loophole"—a period when very few of its natural competitors are on the scene. (A plant's "competitors" are those of roughly similar species.)

Competing plants avoid each other whenever possible. Since plants are rooted to the ground, they cannot just get up and walk away. Therefore they seek to avoid each other in *time* by carefully choosing their period of critical growth. The internal clock of plants enables them to adhere to the complex time schedule which developed over the course of millions of years. This clock notifies them when they have arrived at that time slot between the spring and the fall which is "reserved" for their kind. The internal clock measures the days until they have reached just the proper duration for this particular plant species to bloom.

Astronomical phenomena such as the earth's rotation create time intervals; these intervals in turn profoundly influence biological organisms. The "time clock" mechanism in plants represents just one example of this influence. Seasonal fluctuations in the duration of sunlight affect the life style of all the earth's creatures. Scientists are just beginning to understand the importance of the *temporal* structure of nature.

All living organisms are defined in terms of their spatial attributes: their size, coloring, shape, and conformation. But they are also subject to a temporal order. This order is ultimately founded on the motions of the earth. All creatures pass through cycles of youth, maturity, and aging. We know little of the laws which govern these life cycles. Scientists suspect that further research in this area may give us some clues to the mystery of human aging: why it is that our lives normally span seventy or eighty years, rather than fifty or five hundred.

It would be premature to attempt to deal here with the subject of human aging. Scientists have only just begun to delve into this area. Yet we do know that the time structure determined by astronomical phenomena exercises a profound influence on all living creatures. This influence becomes particularly apparent when the time structure is disrupted by some outside force. We are going to examine two examples of this kind of disruption. The first example involves a species of wild geese which were forced to transfer their summer quarters farther south. The second example will be drawn from human life.

A few years ago Russian zoologists discovered some wild geese in the Baraba steppe region of western Siberia which were behaving as if they had "lost their minds." The birds have their summer nesting grounds in the steppe; each fall they travel south to their winter quarters on the Ganges River about 2,200 miles away. Apart from the mystery of bird migration itself, this whole situation appears entirely "normal." The remarkable thing about these geese lies in their bizarre behavior each fall as they begin their southward journey. The geese always travel the first one hundred miles on foot!

Each August the birds become increasingly restless. One day they all suddenly start heading south. But they do not fly away like "normal" migratory birds. They walk. We must try to imagine a gigantic army of geese—100,000 at the very least—slowly and painfully marching across the steppe like columns of weary soldiers in an army several miles wide. This unnatural form of locomotion completely wears out the geese. They manage to travel only about nine or ten miles a day; and day by day they grow weaker. Soon foxes and other predators move in and decimate the goose army. After about ten days the remnants of the army arrive at a lake-filled area some one hundred miles south of the nesting grounds. The exhausted geese swiftly recover in the water, their natural habitat. A few days later they continue their journey. But now they fly like any other migratory birds. After all they have been through, the remaining 2,100 miles of flight across China and the Himalayas probably seem like child's play.

What can possibly explain the suicidal behavior of these birds? They have fallen victim to the precision of their own internal clocks. The restlessness which causes birds to migrate in the fall is triggered by their internal timepieces. The biological clock measures the length of the days and decrees the exact date when the bird must leave for the south. Under normal conditions this mechanism works with remarkable efficiency. An unusually warm fall might tempt a bird to remain in its summer quarters longer than usual. The bird might then be overcome by a sudden chill on its way south. But the internal clock leaves the bird no choice. As soon as the days shorten to a certain length (a length which varies from one species to another), the bird becomes restless and takes flight willy-nilly toward the south. The same thing happens to the Siberian geese each fall. But in their case the internal clock appears to be telling the wrong time.

The disaster results from the fact that the geese are overcome by restlessness before they have completed their yearly molting: They have not yet acquired their full growth of feathers. Thus they are still unable to fly when the biological clock gives them the order to head south. They cannot simply disobey the order. Therefore the whole army sets off on foot, aiming unerringly for their goal 2,200 miles away. Ten days later their feathers have grown back sufficiently so that they are capable of flight. The difficult journey ahead presents comparatively few complications.

If the geese were only capable of delaying their departure for ten days, everything would proceed smoothly. But the internal clock has fixed the hour when they will grow restless. The same phenomenon which would guarantee their safety under normal conditions betrays them now. For their environment is no longer "normal"; it has slightly altered its character.

Here we have an example of the advantages and disadvantages of instinct as opposed to reasoning power. Some unexplained phenomenon altered the environment of the Siberian geese.

Perhaps the effects of industry developing in the cities invaded their territory in some form; or some change occurred in the climate. In any case, the geese left their native summer grounds and moved a few hundred miles to the south. Their internal clock was not properly "set" to meet the demands of this new locale. In August the northern regions of Siberia remain light much longer than the more southerly regions. (An "eternal day" reigns at the North Pole.) Thus the internal clock of the geese is set to "go off" at a later date in the north than in the south. This clock perfectly adapted the geese to their environment for hundreds of thousands of years. In the unfamiliar environment, the clock reacts as if the year were more advanced than it really is and it forces the geese to leave for the south ten days early. It seems safe to predict that this unfortunate species will not survive the changed conditions for very long.

Human beings are beginning to violate the harmony between the internal clock and environmental rhythms such as the cycle of day and night. To be sure, we do not yet appear to be suffering the tragic fate of the Siberian geese. Nevertheless, we can all come up with a few examples of the discomfort we suffer when we violate cyclical rhythms. A person traveling east or west on a jet rapidly crosses several time zones. He will probably continue to feel tired for about a week; he may also perspire more than usual or suffer from sleeplessness and palpitations. The symptoms of "jet lag" do not result merely from the strain of taking a trip or adjusting to unfamiliar surroundings; people traveling the same distance from north to south—from North to South America, for example—do not suffer from such symptoms.

The discomfort of our jet passenger results from his having crossed several time zones during his flight. (In a flight from north to south, he would cross none of these time zones.) He feels uncomfortable because of the temporary discrepancy between internal and external time. A passenger flying from New

York to Tokyo will land in a time zone which differs by some twelve hours from his internal clock time. He will feel listless as he tries to go about his business. At night when he tries to sleep, he feels wide-awake and gets very hungry. Sometimes it takes the internal clock a whole week to adjust itself to the external rhythm of day and night. Then our traveler perks up again.

This whole situation seems innocuous enough. After all, no one forces us to take trips to Tokyo, and in any case the ill effects are only temporary. But sometimes the conflict between internal and external time seems to have more serious consequences. Our modern industrial society obeys its own rhythms, which often conflict with those of the internal clock and of the natural cycle of day and night. How many of us still eat our meals when we are hungry, rather than when we can fit them into our schedule? Many people eat supper very late in the evening, shortly before they go to bed; breakfast when they are still half-asleep; and lunch whenever their job permits them to go to the canteen or lunch counter. Lunch time is unrelated to individual hunger patterns; it determines mealtimes in terms of shifts. And how many people go to bed when it gets dark? Most of us wait until after the late news on television, or until our shift at the factory permits us to go home. Our life styles are growing increasingly independent of the cycles of nature. One visible symbol of this dissociation from nature is our dependence on artificial lighting, which literally enables us to turn night into day.

Could this dissociation from natural cycles have anything to do with the increasing incidence of nervous disorders in present-day society? For decades now physicians have been noticing how many patients come to them with indeterminate symptoms similar to those of our friend the jet passenger. In fact, doctors claim that they see more patients of this type than those with specific "organic" ailments. And what about the widespread plague of insomnia? Could this complaint be related to the violation of the "natural order" which goes hand in hand with a technological civilization? Perhaps our ability to make the transition from wakefulness to sleep has been impaired by the artificial lighting which characterizes urban life. The German psychiatrist V. E. von Gebsattel once summed up the whole problem by suggesting that it was probably no accident that the sleeping pill was invented at the same time as the light bulb.

No one reading this chapter should assume that we ought now to reverse the course of evolution. Technological society is a one-way street. Few of us would survive the attempt to return to a pretechnological culture. People who talk about "the good old days" generally ignore the other side of the coin. Would they really be willing to pay the price of returning to the past? Would they be able to watch three of their four children die from simple infections? Would they be willing to die themselves from a mere attack of appendicitis? How would they feel about having their teeth extracted without any anesthetic? People are sometimes carried away by romantic enthusiasm.

On the other hand, we should point out that people in the past had a profound sense of the harmony between themselves and nature. This sense of harmony represented far more than the whim of a few romantic enthusiasts or fanatical adherents of natural philosophy. This harmony is a reality—as present-day scientists are beginning to discover. Moreover, those who transgress against this harmony must pay the penalty for doing so.

Thus our well-being partly depends on the relationship between the biological clock and the environmental cycles created by the earth's rotation. The rhythm of day and night provides the most striking example of these environmental cycles. Clearly the natural synchronicity of internal and external time would be gravely disrupted during prolonged space flights. Scientists would also have to take the time factor into account if human beings should ever begin to colonize other planets. Artificial lighting could be used to reproduce the twenty-four-hour cycle during the space flight itself. But how would human beings adapt to the longer or shorter days they would encounter in an alien world?

Probably the human race will never actually colonize other planets. We mention the possibility here simply in order to illustrate the main point of this chapter: Human beings are profoundly influenced by the speed of the earth's rotation. Perhaps it is merely coincidental that our species evolved during a period when the days lasted precisely twenty-four hours. But the twenty-four-hour cycle may have influenced the climate on earth and the structure of living creatures far more profoundly than we now suspect. To be sure, life would probably have developed here even if the days had lasted only thirteen or eighteen hours. But it would not have been *our* earth; and the life on it would have been different from any life we know.

Our Spaceship Solar System

We have noted several ways in which the moon-brake affects all our lives. The twenty-four-hour cycle influences every living creature on earth; grave problems would arise if this cycle were ever disrupted. But when we first began our discussion of the lunar tides, we were actually concerned with quite another problem, the possible source of the earth's magnetic field. It may seem that we have strayed rather far afield; actually our whole argument has just come full circle. We have established the fact that ever since the earth acquired a moon, it ceased to rotate at a uniform speed. This fact argues in favor of the "dynamo theory" of the earth's magnetic field.

We will recall that scientists have devised only one plausible theory to explain why the earth behaves like a giant bar magnet. This theory suggests that the earth's liquid core contains eddies which move independently of the earth's rotation. These independent internal movements enable the core to rotate like the armature of a dynamo, thereby generating electrical current. We know from elementary physics that an electrical current invariably produces a magnetic field.

At this point in our previous discussion of the dynamo theory, we hit a snag. How could an "independent" motion exist in the earth's liquid core? Everyone assumed that the earth had been turning at a uniform speed for billions of years. If this were so, then the earth's core would long since have begun to rotate at the same speed as the mantle. Some scientists tried to rescue themselves from this dilemma by devising the theory of thermal convection currents. These currents supposedly cause whirlpools; the earth's rotation forces the whirlpools to flow in the same direction. Eventually the entire liquid core begins to move as one, thus producing the dynamo effect. We have already noted the flaws in this dynamo theory. In the first place, the discussion of the "smoothing out" of the individual eddies seems overly complex. Second, the theory of thermal convection currents can neither be proved nor disproved. Third, the convection-current theory seems a typical *ad hoc* hypothesis devised to prop up a theory which cannot survive on its own merits. And last, the dynamo theory fails to explain why a planet like Venus, which otherwise resembles the earth in every way, should lack a magnetic field. But once we understand that the earth's rotation has *not* been uniform, all these objections cease to apply. The problem never really existed in the first place!

It is suddenly clear that the earth's liquid core *cannot* be rotating at the same speed as the solid mantle and crust. Tidal friction has been slowing down the earth's rotation. This friction forces the liquid core to "lag behind" the speed of the solid mantle, thus producing a "differential rotation" between the mantle and the core. The earth may well be functioning as a dynamo to produce the magnetic field which shields us from the solar wind. But the "motor" of this dynamo actually lies 239,000 miles from the earth: It is the moon!

At this point it should be emphasized that scientists have not yet reached any firm conclusions regarding the true origin of the magnetic field. They have merely succeeded in eliminating a major objection to the dynamo theory. Only time will tell if this theory is the correct one. Meanwhile, as we have already noted, one advantage of this theory lies in the fact that it explains why the earth's axis of rotation and the axis of its magnetic field should be identical. The rotating earth itself would create the magnetic field as it turned. But it would do so only because the moon is slowing down this rotation. The German poet Hölderlin wrote of the moon as the "pale companion" which soars above us, indifferent to our fate. Hölderlin was mistaken. Without the moon our earth might well be uninhabitable.

Our argument has now come full circle. And yet we have not ended up exactly where we started. Something in our view of the earth has imperceptibly altered. Our planet no longer looks very much like a spaceship. The analogy between our planet and a self-enclosed system like a spaceship simply does not apply. In reality the earth forms only a small part of a spaceship—the cabin containing a whole crew of animals and men. The rest of the spaceship consists of a complex system of cosmic interrelationships suspended in delicate balance. Let us now review what we have learned so far about the role of our earth within this vast network of interdependent forces.

As they journey through space, the members of the "crew" cannot stray too far from the ecosphere of the third planet belonging to a star we call the sun. The term ecosphere refers to that limited area of our planet capable of supporting life. This area may be compared to a paper-thin film covering the surface of the earth. To survive outside the ecosphere, human beings require special equipment such as that used by astronauts and deep-sea divers. Within the ecosphere itself, life is maintained by means of various life cycles. These cycles ultimately depend on forces that lie outside the earth.

Recent research reveals that remote forces in the universe contribute to our survival. The earth is not the stable phenomenon we had always assumed it to be, nor is it isolated from the rest of the universe. The roots of our existence extend deep into interplanetary space. Space itself does not resemble the alien and hostile realm we had pictured for four hundred years.

The sun maintains the earth's life cycles by replenishing its supplies of oxygen, water, and food; the sun also produces light and heat. In order to nourish us, our sun's vast energy supply must be constantly replenished by means of atomic nuclear fusion. These nuclear reactions also release corpuscular radiation —the protons and electrons composing the solar wind. Eventually this wind reaches the edge of our solar system, far past the orbit of Pluto, perhaps 10 billion miles from the earth. (As already stated, we do not know precisely where the magnetic sphere ends and "outer space" begins.) At the edge of the solar system, the solar wind probably comes into conflict with interstellar dust. The clash between solar wind and interstellar matter creates a magnetic sphere which protects us against the cosmic radiation that pours down on us from all the corners of the universe.

The term "space travel" is really a misnomer. We have not yet begun to travel in our own solar system, much less the interstellar space beyond. In one day we could probably walk about twelve miles. Let us imagine that this distance represents the distance between the earth and the moon. On this same scale, Pluto would still be as far away as the actual moon is now! Thus the term "space travel" cannot really be applied to our little excursions to the moon. Moreover, "outer" space actually begins only beyond the shock zone that presumably protects our solar system from the hail of cosmic radiation. We know nothing of conditions in outer space itself.

The gigantic sphere of our solar system may have a total diameter of over 10 billion miles. The outer edge of this sphere represents the outer edge of the spaceship in which we are traveling through the cosmos. The earth supplies only the crew of this ship. Perhaps human beings will never travel outside the spaceship itself, that is, they may never penetrate "outer" space at all. But in another sense our age-old dream of traveling among the stars has already been fulfilled; our entire solar system has been doing just that since it first came into being! A journey among the stars must necessarily take millions of years to complete. How could anything less than a solar system ever last out the journey?

A second invisible magnetic sphere also lends its protection to the earth. This sphere has a diameter of "only" 125,000 miles. Scientists only recently discovered our earth's magnetic field. This field shields the "crew" of the spaceship from the deadly radiation emitted by our ship's atomic reactor, the sun. This second magnetic shield may be created by the moon's effect on the earth's rotation.

Let us now examine the picture we have just drawn of our solar system. We can see the role our earth plays in this picture. But at first glance it may look as if all the other planets were more or less "left out." Do these planets have no influence at all on the ecosphere which all the men, animals, and plants of earth call home? It is too soon to tell. Everything we have just described has been discovered within the past twenty years. Before then, no one even suspected how little we knew about our own earth. We still know very little about the extraterrestrial forces which influence this planet. We have noted that everything in nature has a purpose. It is also true to say that everything in nature has some effect on the outside world. We cannot predict what might happen to us if one of the other planets were to disappear from our solar system. Yet we can safely assume that every corner of our solar system would be affected in some way.

Many discoveries still lie before us. Meanwhile we have learned that our entire solar system plays a role in our survival. Cosmic forces lend support to the frail and improbable organisms which inhabit our earth. We began by pointing out that the earth is not an autarky. Now we can see that in fact our entire solar system forms a united, self-enclosed system—a single entity.

But now another question confronts us: How is our Spaceship Solar System related to the outer space beyond its borders? Does this spaceship travel through the stars in total isolation? Or is it rather touched by influences and forces which reach out from the depths of the cosmos?

Scientists have recently uncovered evidence that a vast network of relationships unites the entire cosmos. Around 700,000 years ago, a "cosmic breakdown" temporarily disrupted the delicate balance of our solar system. The earth's magnetic field completely collapsed! Scientists have now learned that this dramatic event had taken place before.

Journey into the Past

DURING the 1930s a Berlin newspaper published one of the most ingenious April Fool's jokes ever devised. With every appearance of gravity, the columnist announced a sensational new archaeological find: a totally undamaged ancient Egyptian vase. Many plausible details were included in the story. The vase was engraved all over with a sort of spiral design; this design had been cut into the vase before it had left the potter's wheel. Several archaeologists had been involved in the discovery of the vase. One of them had hit on a brilliant notion: He suggested that the spiral line might contain a record of all the sound waves present in the potter's workshop at the moment of its creation—in the same way a phonograph record records sound in its grooves. Scientists had just finished testing this logical-sounding theory and found it correct. They had rotated the vase as it must have been rotated on the potter's wheel in the hour of its creation. The engraved spiral was fitted with an amplifier and the resulting impulse transmitted to a loudspeaker. And behold! Out came an ancient Egyptian folksong! Of course the song was greatly distorted by static. But it was there, as it had been three thousand years earlier when the singing potter carved the finishing touches into the wet clay.

This story has a certain magic, even after one has recognized it as merely an April Fool's joke. The magic stems from the fact that, like every good joke, this tale is based on a genuine insight. In this case the insight consists in the recognition that the past can never be altogether lost. Every event leaves some trace. All these traces of the past add up to form our present. Even a song sung thousands of years ago has not yet altogether disappeared. Air molecules once carried its sound. The mechanical motion of these molecules can never truly be "lost"—no more than any other form of energy. To be sure, the specific pattern once formed by these molecules has long since vanished, and only this specific pattern would enable us to hear the song again. In order to record the song, one would need some way of "fixing" the pattern forever. The story of the vase quite correctly supplied us with a method of doing this.

The past is never really past. True, we can no longer listen to ancient Egyptian folksongs. Nevertheless, in recent decades scientists have been developing new methods of making the past "speak" to us. A short time ago, these methods would have sounded like something out of science fiction. Scientists are now resurrecting events which seemed lost in the earth's remote past. Geophysicists and paleontologists almost seem to be taking a trip through time. Their sciences, which began with the study of sediments and fossils, are becoming almost magical disciplines. They bring back to life creatures and events which seemed to have vanished thousands or even millions of years ago.

By now we have all gotten used to the idea that radioactive isotopes may be used for dating geological phenomena. In the swiftly moving world of modern science, this technique has already begun to look quite venerable. Its basic principle was described shortly after the turn of the century by the brilliant English physicist Ernest Rutherford. It was Rutherford who, in the year 1919, first succeeded in splitting an atom. But he so misjudged the true potential of the atom that shortly before his death he wrote a memorable sentence to the effect that anyone who seriously imagined that the splitting of an atom could ever release substantial quantities of energy was simply fantasizing. This statement was written in 1937, just eight years before the bombing of Hiroshima. But Rutherford was right about one thing. He knew that radioactive elements decay at a uniform rate. Some time before World War I, he recognized the possibility of utilizing this uniform rate of decay as a calendar for dating geological phenomena.

Let us take a look at the phenomenon Rutherford described. Radium and all other "radioactive" elements decay at an absolutely uniform rate; no outside influence can alter this rate in the slightest. The rate of decay among the various elements differs markedly. The term "half life" was adopted to define their various rates of decay. The half life of an element represents the period of time it takes one half the substance of the element to decay. Actually this "decayed" half of the element has simply been transformed into another element.

For example, the half life of radium is 1,580 years. Suppose that we have placed a gram of radium into a sealed container. When we opened the container 1,580 years later, we would find only half a gram of radium; the rest would be lead. Actually radium decays into a whole series of intermediate products before it emerges as lead. These intermediate products are also radioactive; but they have very brief half lives, and after 1,580 years we would have nothing much left but lead, the stable end product of the series. At the end of another 1,580 years the box would contain only one-quarter of a gram of radium, and three-quarters of a gram of lead. Only half the remaining radium would decay during each 1,580 years. And so the whole process would continue for thousands of years until the radium had completely vanished. Long before this it would have dwindled to a quantity we could no longer measure.

Other elements have much longer half lives than radium. It takes thorium no less than 14 billion years to lose half its substance! Other elements have half lives of only a few millionths or billionths of a second. These are the man-made elements, which weigh far more than the heaviest of the natural elements, uranium. The "transuranian" elements cause physicists a lot of grief. It is a bit difficult to study something which vanishes before you have time to take a good look at it!

If a scientist knows how much radium a given mineral once contained, he can calculate when the mineral itself was actually formed. To do so, he need only measure the remaining quantity of radium. He can accomplish the same thing by measuring the quantity of the end product which has developed from the radium decay. The recently deceased German Nobel prize winner Otto Hahn actually succeeded in using radioactive dating methods to determine the age of the earth. Hahn's experiments took place in the 1930s. The radioactive element he worked with was not radium but strontium. Hahn used strontium to date the oldest minerals found in the earth's crust. The age of these minerals would reveal the minimal age of the earth after its liquid crust had hardened. Hahn set this minimal age at 2 billion years. Since that time, still older rocks have been found in the earth's crust. The oldest of these were formed almost 3 billion years ago. Of course the earth's crust remained liquid for a long period before it hardened. Our planet's age is now estimated as at least 4½ billion years.

Scientists are still trying to determine the exact point in time when the earth's crust and the minerals it contains were originally formed. But how do we know what quantity of a radioactive element was originally contained in any given test sample?

Fortunately scientists have devised a few additional tricks which help them to establish a sort of base from which they can begin counting the decay of radioactive material. We are going to examine two of these special methods. Everyone has heard of the famous carbon 14 method. We will also discuss the amazing case of the "geological thermometer"—a method used to determine the temperature of the primeval oceans which covered the earth dozens of millions of years ago.

The carbon 14 method enables scientists to determine accurately the age of organic substances such as bone and plant remains. This method is based on the discovery that the carbon dioxide in the earth's atmosphere contains something besides ordinary carbon; it also contains a very small percentage of a radioactive carbon isotope, which is signified by the chemical symbol C^{14}. Isotopes are atoms of a given element which differ slightly in weight from the normal atoms of this element. Chemically and in all other respects, the isotopes are identical to the normal atoms. Every living creature absorbs some of the carbon isotope C^{14}. This isotope is built into body tissue in the same way as ordinary carbon. But C^{14} is a radioactive isotope; so it slowly decays. Carbon 14 has a half life of 5,600 years. A small percentage of the C^{14} slowly but steadily decays and is eliminated from the organism. Meanwhile this same organism steadily absorbs a fresh supply of the isotope—the plants by breathing the air, the animals by eating the plants. The ingestion of the isotope exactly keeps pace with its rate of decay. Thus the percentages of normal carbon and of C^{14} remain constant in every living organism. Therefore the ratio between the two forms of carbon also remains constant.

Scientists have investigated this constant ratio between the two carbons by studying present-day plants and animals. This ratio establishes the base, or zero point, of the so-called "geological carbon clock." This clock begins to run the moment an

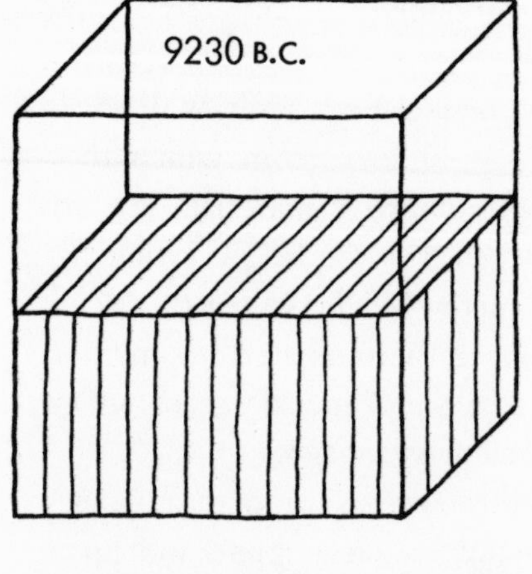

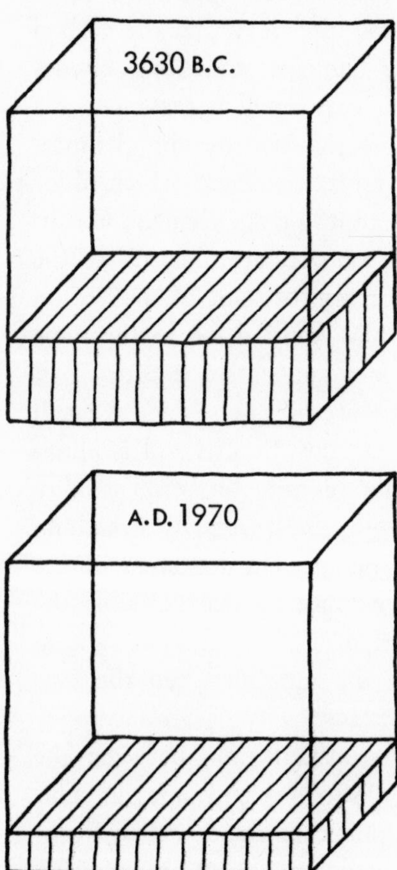

The principle of dating with radioactive isotopes is depicted here, using the example of C^{14}. Carbon 14 has a half life of 5,600 years. The first diagram shows a box half-filled with C^{14} in the year 9230 B.C. *The second diagram shows the same box 5,600 years later; half the original quantity of C^{14} has decayed, leaving the box only one-quarter full. After another 5,600 years (third diagram), only one-eighth of the original quantity remains.*

This diagram illustrates one of the many uses for the method depicted in the preceding diagram. Every living creature contains a specific quantity of C^{14} which remains constant throughout the creature's life. When the organism dies, the C^{14} in its system begins to vanish at the usual rate of decay. The C^{14} content of a skeleton (for example, the skeleton of the mammoth shown here) can be measured. In this way scientists can determine when the animal died. (In these sketches, the decreasing amounts of horizontal shading denote the diminishing quantity of C^{14}.)

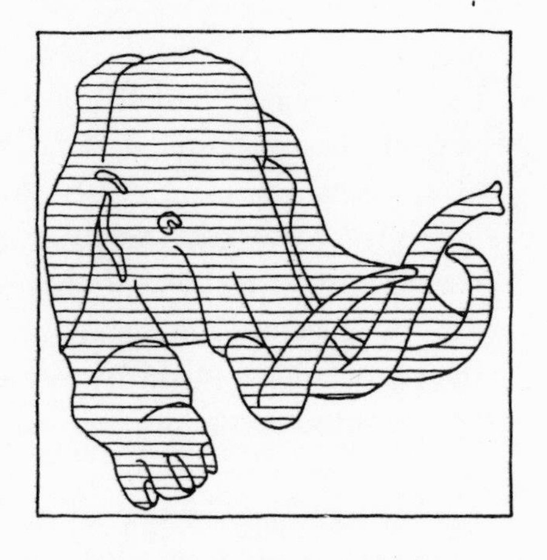

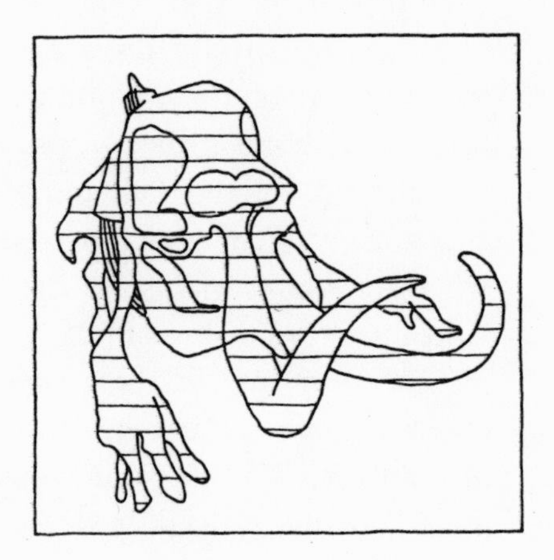

organism dies. At this moment the balance between the ingestion and elimination of C^{14} breaks down. The ratio between the normal carbon and the carbon isotope begins to alter. The dead man, animal, or plant ceases to absorb C^{14}; the C^{14} remaining in the body continues to decay at its usual rate. Meanwhile the normal carbon content of the body remains unchanged. Scientists can measure the degree of alteration in the ratio between the two carbons. In this way they can measure the time which has passed since the death of the sample tissue.

Nowadays if archaeologists discover the remains of a prehistoric campfire in one of their excavations, they just hand over these remains to a physicist. The physicist will examine the bones left over from a Stone Age meal and the half-consumed wood of the campfire in order to determine the ratio between C^{14} and ordinary carbon in these remains. Using the C^{14} method, he can find out when the animals eaten by our ancestors were hunted down and killed; he can also establish when the branches for the fire were cut down.

In theory, our physicist can come within one year of estimating the exact age of these bone and wood remains. In actual practice, such minute quantities of C^{14} are very difficult to measure; this difficulty can somewhat diminish the accuracy of the C^{14} dating procedure. Despite the procedural problems involved in the C^{14} method, it enables scientists to penetrate a past which formerly seemed to have been lost forever. We now know many things we could have learned in no other way. For example, the Lascaux Cave of southern France, world-renowned for the Ice Age cave paintings on its walls, was actually inhabited 15,000 years ago.

Many isotopes besides strontium and C^{14} are now being used as "clocks" to date prehistoric phenomena. But the isotope method has also served as a "geological thermometer." This "thermometer" enables us to determine the temperature of the Atlantic Ocean 50,000,000 or 60,000,000 years ago. Scientists have discovered that snails, mussels, and crabs incorporate two different oxygen isotopes into the calcium molecules which make up their hard shells. These two isotopes can be incorporated into the shell only if the surrounding water maintains a certain temperature. Thus the coexistence of O^{16} and O^{18} in the same shell tells scientists the temperature of the water during the shell's formation—when the living crustaceans were still scrambling around in it. After determining this water temperature, scientists can use the C^{14} method on the same fossils in order to find out their age. In this way they will know at what point in the earth's history the oceans had reached a certain temperature.

Scientists are now attempting to use the annual rings of crustacean shells as dating calendars. These rings are being subjected to microanalysis. Shell rings may enable scientists to determine the sequence of "good" and "bad" summers some 50,000,000 to 100,000,000 years ago. Knowledge of this sequence would help us reconstruct climatic conditions on the surface of our prehistoric earth. Moreover, it would give us some clues as to how our sun behaved millions of years ago. For example, we could find out whether the sun was then subject to the same eleven-year cycle which scientists have observed today.

Scientists are now using a wide variety of methods to make the dead past speak. Recently Dr. H. Dombrowski experimented with some paleozoic bacteria which had been sealed in rock salt for at least 100,000,000 years. The biologist literally succeeded in bringing these bacteria back to life! The tiny prehistoric creatures are now growing and reproducing in the nourishing cultures of our modern laboratories. The metabolic processes of these "long extinct" organisms are being carefully studied. Scientists now have the opportunity to compare the life functions of the prehistoric bacteria with the life functions of their present-day descendants.

Paleontology—the study of the ancestry of life forms now existing on the earth—is no longer restricted to the investigation of bone and fossil remains. Paleontologists are now doing intensive research in one special area of their science; they are investigating the relationships among the species. Their research involves the comparison of protein structures. Hemoglobin and other proteins occur in almost all living creatures. Moreover, these proteins perform roughly the same functions in all creatures (oxygen transport, the breakdown of basic foodstuffs, etc.). And whether they occur in insects, fish, or men, they have a very similar construction. Apparently nature scored a sort of bull's-eye with the invention of these particular proteins, so she continued to employ them over and over. These proteins have existed ever since life first developed on this planet—that is, for more than 3 billion years. Paleontologists today are discovering the basic similarity of all the organisms existing on earth.

But scientists are not encountering only similarities. The proteins which fulfill basic body functions like breathing and metabolization of foods are never altogether identical. Minor variations occur from species to species. These variations occur in those regions of the proteins which are not essential to its biological functioning. Any two given species once belonged to the same family; somewhere in the past they both have a common ancestor. But then their mutual family tree began to branch off, so that various separate species developed from this common ancestor. As time passed, the species grew increasingly dissimilar. Thus the degree of variation between two species depends on how much time has passed since these species first began to diverge. Scientists are now beginning to dream of drawing up a complete "calendar of ancestry." In the next few decades they expect to be able to chart precisely all the significant events in evolutionary history—all the way from primitive one-celled sea creatures down to man himself!

Dating methods like those described in this chapter have already enabled paleontologists to trace much of evolutionary history. For example, we now know that man and the chicken had a common ancestor "only" 280,000,000 years ago. Four hundred and ninety million years have passed since our amphibian ancestors separated from the fish and began to master the land. And a living creature which existed 750,000,000 years ago became the ancestor of all vertebrate life—and of the insects too!

These are only a few isolated examples. Fascinating as these examples are, we must take care not to stray from our real theme. The examples were designed to point out that scientists have devised methods of penetrating the long-dead past. Many people mistakenly believe that our knowledge of prehistoric life has remained sketchy and theoretical. Thanks to the new dating methods, this is no longer the case.

We will now examine one of these dating methods in considerable detail. The method is based on the phenomenon known as "paleomagnetism."

We have noted that scientists today are able to investigate the sun's activity during the age of the dinosaurs. They have also learned the temperature of the seas in which these giant reptiles swam around. Our biochemists can now analyze the metabolic processes of "extinct" microbes. Moreover, for some years now scientists have even been able to measure the properties of the earth's magnetic field during past epochs! The method used to investigate the magnetic field is based on the phenomenon of "fossilized magnetism." This exceptionally weak form of magnetism is very difficult to measure, but the basic procedure is simplicity itself.

Many minerals in the earth's crust contain iron; such rocks are very readily magnetized. Volcanic rocks may contain large quantities of iron. Shortly after World War II geophysicists hit on a novel idea. One hundred million years ago an active volcano may have erupted somewhere on the earth, flooding the surrounding area with boiling lava. As long as the lava remained hot, the iron salts it contained displayed no magnetic properties. (Above a temperature of 770°C iron has no magnetic properties.) But as soon as the lava reached the earth's surface, it began to cool; in time it fell below the critical temperature level. The iron compounds in the hardening rock regained their magnetic properties—under the influence of the earth's magnetic field!

In other words, the iron in the lava is magnetized along a north-south axis which corresponds to the poles of the earth's magnetism. Millions of years after the actual volcanic eruption, this magnetic polarization can still be detected in rocks. It constitutes a genuine "fossil." Scientists can enter a volcanic region and cut through the rock layers until they reach the oldest layer of that region. In our sample case, the volcano first erupted 100,000,000 years ago. The magnetism "frozen" into this layer can now be measured. In this way scientists can find the direction of the lines of force in the earth's magnetic field 100,000,000 years ago. In order to date the lava layer itself, they will use the isotope method already described.

Once it had been thoroughly tested and documented, this procedure came into general use. Scientists began to seek out and measure remains of "fossilized" magnetism all over the earth. Rock layers of widely varying ages were examined. At first scientists devoted most of their attention to the technical problems involved in measuring such a weak form of magnetism. (The magnetic fields of these rocks are only one percent as strong as the earth's magnetic field.) At this point, no one expected any remarkable developments. But suddenly geologists were confronted with the two most exciting discoveries made in their field during the past twenty years.

Geophysicists had always assumed that in the past the earth's magnetic field must have roughly corresponded to the field we know today. That is, it must always have followed its present north-south orientation. Thus geophysicists supposed that all the paleomagnetic samples they gathered would yield more or less identical results. They had a number of reasons for believing that the magnetic field could never have shifted more than a slight degree. For one thing, they knew that the axis of the earth's magnetic field was closely related to its axis of rotation. Throughout the history of the earth, this axis of rotation had never markedly shifted its position. To be sure, there is a phenomenon known as the "wandering of the poles." From time to time the earth's axis clearly executes a sort of circular motion. But the axis shifts by no more than some thirty-two feet a year.

A small experiment can help us to understand why the earth's axis never shifts by more than a few feet. Let us take a large heavy humming top of the sort we used to play with as children, and try to shift the position of the spinning top by pressing on its upper axis with one finger. We will encounter resistance; the top refuses to budge! Most people are surprised to see how strongly a spinning object resists any attempt to alter its position. We live in a world in which everything is in motion. In such a world, the most stable form of all is a rotating platform. A spinning platform remains independent of outside influences. It therefore provides a "point of rest" from which to view the world and calculate one's course. For this reason, the automatic guidance systems of our modern space rockets function by means of rotating platforms. The stability of these platforms is maintained by the rocket computer, which periodically checks out the position of the platforms in relation to certain stars. (Stars may be used for guidance purposes because they are so far away that from our perspective they never appear to move.) The computer then makes any necessary adjustments in the position of the rotating platforms.

Thus a rotating object resists any outside force which attempts to shift its axis of rotation. This force of resistance actually represents a force of inertia. If this force were less powerful, the paper-thin crust of our planet would long since have been torn to shreds; the earth's glowing core would have hardened into countless isolated drops of matter in the ice-cold universe of space. But so far our earth has not encountered a force powerful enough to shift its axis by more than a few feet a year.

We have noted that the axis of the earth's magnetic field is closely aligned with its axis of rotation. Since this axis of rotation has never moved very far, scientists assumed that the axis of the geomagnetic field had also stayed "close to home." Therefore it seemed inconceivable that paleomagnetic rocks would reveal anything more remarkable than the familiar north-south polarity of the modern magnetic field. But scientists had a surprise in store for them—or rather, two surprises. They found two separate deviations from the "traditional" sort of magnetism. The deeper they delved into the past, the more traces they found of the two "paleomagnetic anomalies."

Scientists were completely bewildered by the first type of anomaly they encountered. At first they attempted to write off their findings as simple errors in measurement. (The weak "fossilized" magnetism had in fact proved very difficult to measure.) None of them could make head or tail of the contradictory data they were recording; their measurements seemed to fluctuate in a wholly capricious way. At times the lines of magnetic force deviated sharply from the modern north-south pole of the earth; at other times the deviation was slight. But this was not the worst of it. When geophysicists began to compare rocks from different continents but from *the same geological epoch*, the measurements from the various continents kept contradicting each other. For example, scientists might test volcanic rocks in America which were 200,000,000 years old; these rocks would indicate that the North Pole at that time lay in the area of modern Siberia. Then geophysicists would compare these rocks with European samples from the same period. But the European samples revealed that the North Pole must have been situated near southern Greenland!

Geophysicists could not explain these confusing data. But every good scientist must possess a large measure of patience and tenacity. The geophysicists continued to assemble data from various geological epochs and entered the data in their charts. A few years later their patience was rewarded. In order to recognize the picture formed by a mosaic, one must first have placed most of

The eastern coast of South America has a contour which "fits" into the coast of West Africa. This similarity of contour tipped off scientists to the phenomenon of continental drift. In recent years many incontrovertible proofs have been found that this drift actually occurred and is still going on today. The dotted lines show how the two continents were situated hundreds of millions of years ago.

the tiles in their proper positions. In the same way, the data relating to fossilized magnetism slowly began to take shape.

It stood to reason that the earth could have only one North Pole at any given time. By the same token, it could have only one South Pole, which must lie directly opposite the North Pole. Granting this assumption, all the data which had been assembled could have only one possible explanation. The continents had been continually shifting position on the earth's surface! The geophysicists conducting the paleomagnetic investigations ended by accomplishing something they had never dreamed of: They proved the theory of "continental drift." German geophysicist

Alfred Wegener had already advanced this theory in 1912, but at that time most scientists had rejected it.

One of Wegener's proofs in support of his theory had been the apparent similarity in contour between the eastern coastline of South America and the western coast of Africa: "The knee of South America fits into Africa's groin." A quick glance at a world map reveals that the two continents actually fit together like two pieces of a puzzle. Before Wegener, countless scientists had noticed these matching contours without giving them a second thought. But Wegener decided that the similarity represented more than mere coincidence. He made the bold assumption that the two continents must long ago have formed a single giant continent which had then split apart. Once this split occurred, the two continents continued to drift apart, moving along a viscous layer of the earth's outer crust. Wegener patiently continued investigating and comparing geological specimens. In time he discovered additional, far less obvious pieces of the "continental puzzle." For example, he discovered that the coastline of India fit quite snugly into the southeastern coast of Africa.

For a time the theory of continental drift was widely discussed. But most scientists soon rejected it. They maintained that no one, including Wegener himself, could explain what force had set whole continents to drifting around the surface of the earth. Nowadays most geophysicists accept Wegener's theory as a fact. Some unknown force compels the continents to move about in much the same way that huge ice floes moved across the continents during the Ice Age. This unknown force moves the continents in slow motion, at a rate of only a few inches a year.

Nowadays no one doubts that the continents move; the question is, what moves them? Many geophysicists assume that areas of varying temperatures inside the earth's crust shift position, thus producing convection currents. We may recall that scientists made much the same assumption with regard to the magnetic field. The convection currents which cause continental drift must lie much nearer the earth's surface than those currents which may help to produce the magnetic field. Moreover, these surface currents must move much more slowly than currents in the earth's liquid core.

One recent theory suggests that the continents are being driven apart by the pressure of lava spurting from fissures in the ocean floor. A number of these huge volcanic fissures, which continually emit streams of boiling lava, have been discovered in the floors of all the major oceans. But the question of what causes continental drift remains undecided.

Wegener did not live to see his theory take hold. In 1930 he participated in an expedition to Greenland, where he died under rather mysterious circumstances. At that time he was fifty years old. It was rumored that his mental stability had suffered acutely from the zeal with which his colleagues had attacked his theories. Some people still suspect that his death may actually have been a suicide.

Scientists slowly gathered evidence substantiating the theory of continental drift. Shortly after World War II geologists found proof that the coasts of Africa and South America must at one time have formed a single mass. Rock samples were taken from the "matching" regions of the two coasts. These rocks clearly came from the same formation! Wegener's theory was still not totally accepted, but various scientists patiently continued to test it out. They assembled more and more evidence in its favor. We shall examine only one of these many proofs. This particular piece of evidence shows what scientists can accomplish when they do not restrict themselves too narrowly to their own field of research.

Surprisingly enough, a discovery in zoology made a major contribution to proving the theory of continental drift. In 1968 zoologists discovered a microscopic species of crab in the sediments of the Amazon Delta. Certain physical hallmarks identify this crab as a member of a particularly ancient species. In fact it represents a genuine "living relic" or "living fossil"—the remains of a prehistoric species. A number of such prehistoric species have somehow managed to survive in small isolated "oases" in various parts of the world.

The zoologists set about the business of classifying their prehistoric South American crab. In order to locate its closest relatives, they compared it with other crab species which had already been classified. They experienced a startling revelation: The crabs were not quite as "unknown" a species as the zoologists had first assumed. These creatures were so similar to another crab species that scientists could scarcely tell them apart. But this second, related crab species existed only on the other side of the ocean—in the ground water and silt layers of a few West African river deltas! This amazing resemblance could not be explained as a mere coincidence. Zoologists ruled out the possibility that either species could have migrated across the Atlantic to the opposite coast. Both were fresh-water species which could not have survived a journey through the ocean. The zoologists drew the only possible conclusion: The two crustacean species must once have rubbed shoulders in the same corner of the earth. Thus the two continents must originally have been joined.

During the past few years, paleomagnetic research has lent additional support to Wegener's theory. We have noted that geophysicists painstakingly assembled all data relating to fossilized magnetism. They systematically charted all the rock samples gathered from each volcanic region, classifying the samples according to their age. Scientists first tested the deepest (that is, the oldest) rock layer in each region; they recorded the direction of the lines of magnetic force contained in these oldest rock samples. They then recorded the magnetic direction of the second oldest layer, then of the third, and so on.

After a time, a clear picture began to emerge. The various magnetic data revealed a kind of gradual "drifting" pattern. The magnetic polarization of the oldest rock layers deviated sharply from the north-south line formed by our present poles. The magnetic poles of more recent rock layers gradually approached

The recently discovered phenomenon of paleomagnetism enabled scientists to reconstruct the drift of continents and islands. This drifting movement has continued throughout the earth's history. For example, during various geological epochs, a certain volcano might erupt many times and cover the land with many successive layers of lava. At the same time, the land mass to which the volcano belongs will be gradually shifting position. (In the diagram a curved arrow indicates the direction of this shifting motion.) When scientists examine the various layers of lava, they find that each layer contains a different magnetic orientation. During the intervals which separate the volcanic eruptions, the volcanic land mass alters its position. Thus the lines of magnetic force "frozen" into each layer of lava deviate somewhat from the line of our present north-south pole. The degree of deviation depends on when the volcanic eruptions occurred; lava layers formed from very ancient eruptions deviate most sharply from the present north-south pole. This is true because the volcanic land mass has been continually shifting ever since these ancient eruptions occurred.

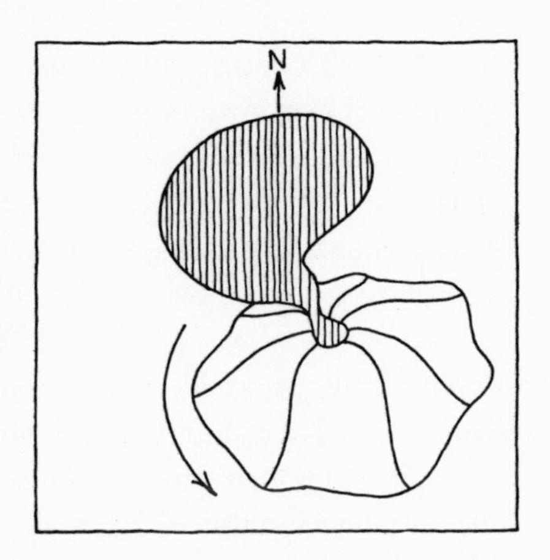

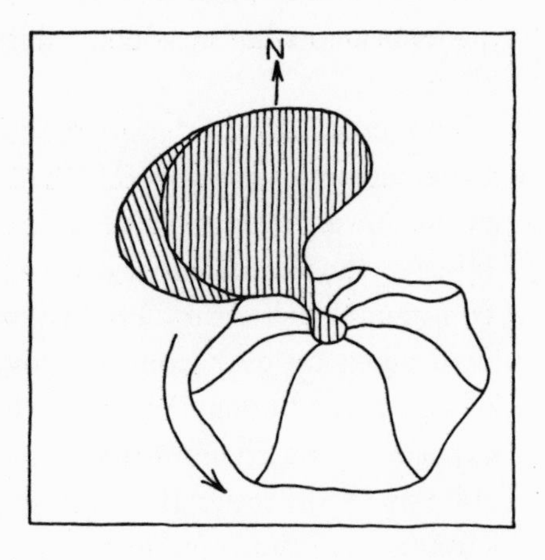

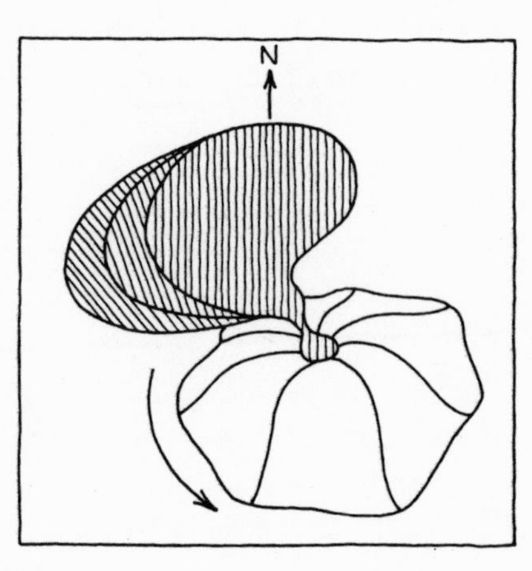

our modern compass directions. The most recent layers corresponded almost exactly to the line of our present poles. Geophysicists examined rock samples taken from within a single continent; in all these samples, the "drifting" of the poles seemed to follow the same pattern. But this intracontinental pattern differed markedly from the patterns of all the other continents.

Clearly the North and South Poles could not have maintained several contradictory positions at one and the same time. Nor could they have shifted at constantly varying speeds throughout a single geological epoch. Thus the thousands of painstaking measurements gathered from all corners of the world left room for only one conclusion: Over the course of millions of years, the continents had gradually shifted their positions in relation to the poles.

This conclusion did more than simply verify Wegener's theory. Paleomagnetic investigation had also revealed the sequence of events in various regions of the earth. That is, scientists could now observe the actual historical unfolding of the phenomenon of continental drift; they could observe its tempo and the precise path taken by each continent over the course of many millions of years. We human beings are short-lived; we can never actually experience the changes which slowly take place over inconceivable spans of time. But thanks to the new methods of measurement, paleomagnetic researchers almost received a taste of immortality. It was as if they had just been handed an accelerated film version of the history of the earth's crust.

Disaster Strikes Our Magnetic Shield

WE HAVE seen how the magnetism "frozen" into ancient rock layers slowly altered its magnetic orientation. Each continent had followed its own pattern of "magnetic drift." Geophysicists discovered this type of magnetic anomaly while studying the phenomenon of paleomagnetism. But a second surprise lay in store for them—this one even more startling than the first. In fact, this new discovery almost seemed to contradict their earlier findings. Scientists had detected a *gradual* "drifting" pattern in most of their rock specimens. But now they began to encounter instances of a very *abrupt* shifting in the direction of lines of magnetic force. The deeper geophysicists dug into the past, the more examples they found of this second form of magnetic anomaly. The sudden shifts in the direction of the lines of magnetic force seemed to have occurred at widely spaced intervals—intervals of from 100,000 to 1,000,000 years. Moreover, rock specimens of the same age all showed the same shift of magnetic direction—no matter where they came from on the earth!

This second discovery seemed even more mysterious than the first. Even now it has not been fully explained. Paleomagnetic findings indicate that this shifting in the direction of the lines of force in the magnetic field must have occurred simultaneously all over the earth. Moreover, the degree of shift was always identical—exactly 180°. Thus each time the phenomenon occurred, the two poles of the earth's magnetic field must have been completely reversed! In other words, our present-day geographic North Pole has not always been the North Pole of our magnetic field. In fact, the North Pole has been "north" for a relatively short time—for only about 700,000 years. Before this time it was the "South" Pole. If we could travel back into this earlier epoch with a compass, then our compass needle would be pointing south. At that time the true North Pole lay in Antarctica. But this situation persisted for only 300,000 years. Before this time, the North Pole was identical with our present North Pole—at least for 100,000 years. For almost a million years before that, the South Pole had been our magnetic north. And so the pattern continues as we penetrate deeper into the past. The earth's magnetic field has continued to reverse its poles at intervals of 100,000 to 1,000,000 years.

Paleomagnetic data conclusively proved that this dipolar reversal of the magnetic field had actually taken place. But at first no one could even begin to explain the phenomenon. Scientists were baffled as to its causes, and they wondered even more about its effects. How had the repeated reversals of the earth's magnetic poles affected living conditions on the surface of our planet?

Once again we have encountered an instance of our basic theme, the interrelatedness of all cosmic phenomena. When geophysicists first discovered the phenomenon of dipolar reversal, they already knew certain basic facts about the earth's magnetic field. They were familiar with the Van Allen radiation belts; they also knew how the earth's magnetosphere shields us from the harmful radiation of the solar wind. Thus scientists at once began to wonder how the dipolar reversals had affected the radiation belts and the entire magnetosphere. And more importantly, what had become of the living creatures on the earth's surface?

In our description of a dipolar reversal, it sounded as if the whole process occurred almost instantaneously—as suddenly as a lamp can be switched on or off. But the magnetosphere surrounding our planet covers a vast area. Moreover, huge masses in the earth's core are involved in the production of the magnetosphere. No change capable of affecting these vast regions could take place overnight. In fact, ephemeral creatures like ourselves would undoubtedly regard a dipolar reversal as anything but "abrupt."

For some reason, the earth's interior must periodically cease to function as a dynamo. Eventually the earth renews its dynamo activity once more, so that the magnetic field is gradually restored. Scientists are not certain what factors trigger this whole process. However, at the moment we must assume that the geomagnetic field has actually collapsed *twice* as often as our present data indicate. Paleomagnetic data record only the actual, completed dipolar reversals of the magnetic field. That is, these data do *not* record those times when the magnetic field breaks down and then is restored *along its previous poles.*

The laws of probability suggest that the collapsed magnetic field has not always been restored along the *opposite* poles. Actually, there has always been a fifty-fifty chance that the North Pole would be restored to the region of the earth it had occupied before the collapse. Thus the earth has probably lost the protection of its magnetic shield twice as often as our present data reveal.

How long did these "unprotected" phases last? Estimates vary widely. There are still many gaps in our knowledge of how the magnetic field is created in the first place. Scientists believe that each time the field collapses, the earth must remain without protection for *at least* a thousand years. During these periods, all the living creatures on our planet are exposed to the increased intensity of the solar wind. Human beings think of a thousand years as an enormous stretch of time. Nevertheless, geophysicists are justified in regarding the process of dipolar reversal as "abrupt." After all, these thousand-year stretches are separated by intervals of up to a million years. Throughout these long intervals, the magnetic field appears to remain quite stable. Compared with its periods of stability, its periods of collapse seem to pass in the twinkling of an eye.

Many geophysicists believe that the magnetic field remains "out of order" for intervals much longer than a thousand years. Their estimates have ranged from 2,000 to 10,000 years; but most scientists now agree on a span of at least 2,000 to 5,000 years. In any case, periodically the earth has been exposed to the solar wind for periods of *at least* a millennium. What happened to our planet during these sieges of radiation?

Scientists have been debating this subject for many years. During the past two years, new data have come to light which seem to have brought the long debate to a close. Until this time, scientists had been divided into two camps. One group maintained that the solar wind must have profoundly affected all life forms on the earth's surface; in fact, entire species might have died out from the effects of solar radiation. We know that in the course of evolutionary history, a number of species have become extinct. Frequently scientists have been unable to explain why these particular species should have died out. When paleontologists learned of the periodic breakdowns of the earth's magnetic shield, they believed they had found a possible cause for the inexplicable extinction of certain species.

A second group of scientists felt that their opponents had painted too black a picture. These scientists maintained that the temporary collapse of the magnetosphere had had virtually no effect on living organisms. One factor clearly argued in favor of this "optimistic" view, the nature of the earth's atmosphere. We have noted that the moon has no atmosphere at all and is therefore completely exposed to the solar wind. But the earth actually has *two* protective shields, the magnetosphere *and* the earth's atmosphere itself. To be sure, the particles of the solar wind constitute a highly intense form of radiation. But the air layer above the earth extends into the sky for dozens of miles; it acts as a cushion to absorb the solar particles. Scientists determined the density, altitude, and composition of the earth's atmosphere; they also analyzed the density, composition, and speed of the solar wind. By correlating the two sets of data, scientists computed the mathematical probability that a single particle of solar wind might succeed in penetrating our defenses. The solar wind hurtles toward us at a thousand times the speed of sound. Yet the chances are slight that any of its particles will ever reach the earth's surface. The phenomenon of the polar lights proves this fact. The polar lights appear at an altitude of fifty miles and more above the earth. These lights are produced by the clash between solar wind particles and the air of our upper atmosphere. We have already noted that the polar regions have no magnetic shield; that is, in these regions nothing but the atmosphere itself shields the earth from the solar wind. The polar lights mark the point at which the protons and electrons of the solar wind are halted by the earth's "air cushion." Most of the particles never come within fifty miles of the earth's surface.

It might seem that these facts about the earth's atmosphere should have settled the scientific debate over the probable effects of corpuscular radiation on living organisms. Moreover, it must appear that our earlier chapter on the earth's magnetosphere greatly exaggerated its protective function. Unfortunately, the earth's relationship with the solar wind is not quite this clear-cut. We have noted the opinion of some scientists that solar radiation had seriously damaged the earth's organic life. But even these scientists never assumed that the radiation killed its victims *directly*. In other words, the solar wind does not constitute some sort of deadly ray which strikes down everything it touches. The true danger posed by the solar wind lies in the phenomenon known as "secondary radiation." Many scientists believed this radiation capable of disrupting the normal course of biological evolution on our planet. In other words, not a single living creature would actually *die* from its effects, but entire species might suddenly be in danger of extinction. In the following chapter, we will see how radiation has affected life on earth.

The Motor of Evolution

WE HAVE noted that the carbon isotope C^{14} represents a variation on the ordinary carbon atom. Carbon 14 differs from normal carbon in one respect only: It is a radioactive element. That is, C^{14} continually emits radiation, thus slowly diminishing its substance at an absolutely uniform rate. Up to now we have been concerned only with its uniform rate of decay. Carbon 14 provides scientists with a built-in "precision timer" which may be used to date ancient organic tissue. All living tissue contains carbon. Therefore C^{14} may be used to date bones and wood from the Stone Age. However, we have not yet considered the question of what produces C^{14} in the first place. We know that C^{14} maintains a constant level in the air we breathe. Yet we also know that it "decays" (throws off radiation) at a constant rate. Therefore something must be constantly replenishing the supply of C^{14} in the earth's atmosphere.

Scientists now understand what produces our constant supply of C^{14}. This carbon isotope descends to us from the uppermost layers of the stratosphere. Cosmic radiation and particles of the solar wind continually slip past the protective walls of our radiation belts. These two kinds of radiation steadily bombard the normal carbon of our upper atmosphere some sixty miles and more above the earth. The radiant particles themselves are captured by the earth's atmosphere, so that they hardly ever reach the earth. But meanwhile the particles strike our atmosphere with great force. The force of impact transforms the normal carbon atoms into the radioactive isotope C^{14}. Thus the solar particles themselves hardly ever succeed in penetrating our "air cushion" in order to reach the earth's surface. But the "secondary" product C^{14} becomes uniformly diffused throughout our atmosphere. It is absorbed by all our plants and travels by way of the food chain into the bodies of animals and men.

Let us assume that we have shielded ourselves from a hail of bullets by hiding behind a thick armor plate. This plate protects us from the bullets themselves. But as the shots strike the plate, they break off small chips of metal which ricochet in all directions. We are now in danger of being hit by one of these small chips. In the same way, C^{14} is produced by the impact of radiation particles striking the carbon of the earth's upper atmosphere; and we are continually being "hit" by it. Human beings absorb this C^{14} into their tissues at a constant rate, as do all other living organisms. So constant is this rate of absorption that it can actually serve as a paleontological "clock" to date past events.

Exactly how does C^{14}, this secondary product of solar radiation, affect the human body?

People today hear a great deal about the "peaceful" uses of atomic energy. Nevertheless, words like "nuclear fission" and "radioactivity" tend to have a purely negative connotation. We all fear the mighty force of nature—the atom—which has been unleashed by modern technology. Thus it may surprise many of us to learn that the continual bombardment of C^{14} plays an indispensable role in the lives of all the higher organisms on earth. If the earth's atmosphere did not contain a certain percentage of C^{14} and of other radioactive isotopes, we would probably not be alive today! The same statement applies to all the other higher animals and even to all the plants. To be sure, C^{14} plays no essential role in our metabolic processes. All animals and plants now living could survive perfectly well if their supply of C^{14} were suddenly cut off. If we were all transported to another planet whose atmosphere was totally radiation-free, we would continue to live out our natural lives.

The role played by C^{14} and other radioactive elements in our atmosphere is played behind the scenes. These elements create the phenomenon known as "background radiation." This radiation has long been a permanent element of our earth environment. It apparently acts as a sort of accelerator to regulate the speed of the evolution of life. In other words, it determines the rate at which existing species develop and new species come into being. We could survive without the effects of radiation, but if this radiation had not existed, our species would not exist either. For this radiation started the motor of evolution running. Lacking the effects of radiation, life might not yet have developed beyond its initial stages. Life began three billion years ago. Yet today the earth might still have been almost uninhabited; it might have contained nothing more than primitive, unconscious one-celled organisms drifting passively through its oceans.

In order to understand why radiation should have played such a major role in the evolution of life, we must now examine the basic principles underlying evolution itself. Only then can we appreciate how life on earth has been affected by the repeated reversals of polarity which disrupted the earth's magnetic field.

Life itself is an improbable phenomenon. It literally contradicts the laws of probability. The odds against the existence of any form of life whatsoever are simply astronomical. Many complex processes unite to maintain the life of a single "primitive" cell. We have not yet discovered many of these basic life functions; nor do we thoroughly understand those we have discovered. Inconceivably vast forces join in the task of producing a single organism; then they must struggle to keep it alive for the briefest of spans before it falls back into the state of inanimate matter. Anyone who thinks about all this for a few minutes gets dizzy from wondering how it all came about.

The question has no ready answer—at least not for the present. Many people are tempted to apply the word "miracle" to anything they do not understand. This word connotes a phenomenon which by its very nature defies scientific explanation, which surpasses all human understanding. Unlike "normal" phenomena, miracles are regarded as violating the laws of nature. Moreover, they are assumed to be manifestations of some form of "supernatural" force.

When we think about the phenomenon of life, it seems easy to conclude that this phenomenon is indeed "miraculous." All the same, we ought not to cheapen the word miracle by applying it to anything we happen not to understand at the moment. For centuries now Christianity has waged a tragic and unnecessary war against Western science. The theologians were defeated time and time again; but they brought these defeats on themselves. They irrationally persisted in basing their concept of the miraculous on the phenomenon of human ignorance. If I call everything I do not understand a "miracle," I run the risk of having to give up my miracles one after another; for meanwhile other people will have succeeded in explaining them.

It is clearly dangerous to define as a miracle one's own lack of comprehension. At the very least, this procedure represents a "tactical" error. Moreover, this is a superficial approach to a complex problem. People often speak as if any given situation had to be *either* "intelligible" *or* "miraculous." But why must these concepts be mutually exclusive? Why should I cease to find something miraculous simply because I have in some sense "understood" it? Many people greatly overestimate the power of the human mind. Whenever scientists solve one problem, they discover the multitude of other problems that had been hiding behind it. Each answer leads to new questions. The world as a whole completely transcends our poor efforts to understand it.

But now let us return to our original subject—the complexity of even the most "primitive" life forms. Nature expends great effort in maintaining the lives of even the most modest of her creatures. She improves a species in order to increase its chances of survival; and she tenaciously clings to every detail of construction once it has proved its worth.

At one time or another most of us have drunk orange juice to build up our resistance to colds; or we drank hot lemonade to help ourselves get over a cold we already had. The vitamin C contained in oranges and lemons illustrates how we all benefit from the tenacity of nature—her tendency to "memorize" successful details of construction. How does it happen that a substance produced by plants can prove so beneficial to the human system? The citrus tree synthesizes vitamin C for its own benefit, not for ours. Yet human beings require this substance for survival; and our bodies are incapable of producing it themselves. How is it that the lemon can supply vitamin C in exactly the form our bodies need? A vitamin does not serve as a food or any form of metabolic "fuel." Instead, it is a complex molecule whose form "fits" into certain parts of a living cell as precisely as a key fits into a particular lock. This molecule triggers certain essential body functions which cannot take place when the body is suffering from "vitamin deficiency."

How did a "key" which fits our body cells come to be inside a citrus fruit? Vitamin C exists in the citrus fruit and in many other plants. This fact represents one of many proofs that all terrestrial life developed from a single source: every living organism on the earth today is related to every other. Human beings are very distant cousins of the citrus tree. Once nature had invented the "key" we call vitamin C, she used this successful invention over and over. Today this key fits the cells of the most widely varied creatures—creatures we no longer think of as being related. This fact proves that vitamin C must have been invented rather early in the course of evolution; otherwise it could not now play a role in the metabolic functions of so many varied organisms.

We are all aware of this basic "tenacity" or conservatism of nature. But for the most part we know it by another name—"heredity." Heredity may be defined as that biological principle which insures the preservation of certain traits in future generations of a species. The resemblance between a child and his father expresses this basic form of tenacity—the tendency of nature to reduplicate anything which has once been achieved. If nature did not tend to reduplicate forms, then each individual creature which came into the world would have to be separately created; all its life processes and indispensable organs would have to be created from the bottom up.

But a child resembles his mother as well as his father. He may also look a little like his grandparents. As he grows up, his parents discover the features of various other relatives in the child's changing face. Yet despite all these resemblances, his face is still his own—something brand new, an individual physiognomy which never existed before.

No doubt this phenomenon strikes us as rather banal. Yet in the "obvious" phenomenon of a child's face we can read nature's second solution to her problem: how to evolve new and more highly organized species. The conservative principle of preserving what had already been achieved was not enough. An absolutely "conservative" nature would simply have gone on exactly reproducing each generation in the next. All the generations would then have been identical. This process would not have permitted any species to develop. Life on earth would never have had a "history" at all. No conscious life would have existed to the end of time. Nothing would ever have lived but the senseless repetitions of a single primitive organism—an organism which one day in prehistoric time hit on the knack of producing another living creature like itself.*

If "evolution" were ever to occur, nature had to strike a balance between two conflicting interests, the preservation of older forms and the possibility of new ones. How strange to realize that the conflict between progress and tradition should have been posed so early and on so fundamental a plane! Absolute continuity would have excluded change; but radical change could have destroyed what had already been accomplished. Charles Darwin achieved a number of amazing things. One of his major achievements lies in his clear recognition of the basic conflict underlying the concept of evolution, the conflict between tradition and progress. Darwin lived in an age when genes and chromosomes were still unknown. Yet he recognized nature's solution to the conflict between stability and change. This solution lay in the random or spontaneous variation of hereditary traits.

These variations in inherited traits are known as "mutations." Some years ago we discovered how these mutations occur. Every nucleus of every cell in every living organism contains genes. These genes carry all the hereditary traits which distinguish this particular organism. The genes are composed of molecules; the composition of these molecules determines the individual traits. In school we learned about the complex process of cell division. This mechanism ensures that each time the cell divides, those parts of it which contain the genes will reproduce themselves with the utmost precision. Thus each of the two new cells will acquire exactly the same set of genes. The division of the cell nucleus directly transmits hereditary traits. The process of

* This is a purely hypothetical example; no living species could ever *exactly* reproduce itself. Environmental factors would prevent this from occurring. Every organism depends for survival on a reciprocal relationship with its environment and with members of other species. Thus no species could reproduce its original traits for any length of time. Environmental factors would soon intervene, inhibiting some traits and promoting others.

transmitting genes through cell division reveals how nature solved the first of her two problems, the problem of preserving an advantageous trait down through the generations.

We also know how nature solved her second problem, the problem of change. Darwin intuitively predicted what we have now discovered to be a fact. In a small percentage of cases, something goes wrong during the complex process of cell division. Accidental errors may occur. One of the molecules bearing hereditary traits may fail to divide quite correctly. Or again, the cell may already have divided; each half begins to make itself into a whole cell. At this point one molecule may "inadvertently" pick up the wrong particle. Or the right particle may end up in the wrong part of the molecule. Many other such small errors can occur. The result is the creation of a gene which is not *absolutely* identical to the corresponding gene of the original parent cell. A gene is born which enables the organism to pass on a novel hereditary trait. We refer to such a genetic variation as a "mutation."

Mutations function more or less as the "motor" of evolution. Mutations represent an organism's only hope of changing over the course of time. These wholly capricious hereditary "leaps" give a plant or animal species something very precious, the chance to adapt more successfully to a climatic change or some other alteration in its environment. Or the species may simply improve its chances of survival in its old environment.

This opportunity to increase the efficiency or adaptability of an organism arises wholly by chance. Some people still refuse to believe that complex organisms may develop through random mutations. But geneticists have long been convinced that the molecular variation of hereditary traits offers the only hope of change for any species. Only by means of mutations can a species adapt to a changing environment. In other words, mutations are the source of that "flexibility" which each species reveals in the course of its development. Certain species reveal a far greater degree of flexibility than others.

At this point some people would argue that the many complex structures composing a living organism could never have developed through a process of random mutation. Living creatures are orderly structures, not manifestations of chance. Impressive as this argument may sound, it does not really hold water. It will not be possible to devote much space to discussing or refuting this point of view, but there are a few factors which opponents of the random mutation theory often overlook. These items may be included here because they directly relate to our basic discussion.

In reality, mutations by themselves do not further the evolution of a species. Yet mutations *do* furnish the absolutely indispensable "raw material" of progress. This raw material still has to be shaped by the environment. Of the many mutations which occur in nature, the environment "selects" only a few. Every environment constitutes an ordered structure. All nature is ruled by temperature gradients and rhythmic cycles like the cycle of day and night. Moreover, every environment contains living creatures which function as prey, as partner, or as predator to every other. All these organisms engage in various forms of purposeful behavior. While a mutation may be "random," an environment is not.

The environment introduces the element of order into the evolutionary scheme. Many forces determine whether or not a particular genetic mutation will be accepted as a permanent trait of the species. In order to pass on its mutant trait, an organism must be able to survive in its environment. It must also be able to reproduce and to raise its young, so that the mutant gene may be preserved. The phrase "natural selection" refers to the fact that the environment favors only those mutations which aid a species to survive. In other words, in this case order can arise "by chance." Only those mutations survive which "accidentally" happen to conform to an external order.

Thus the earth does not function simply as the stage on which the evolution of life is acted out. The earth is also one of the characters in the play. The earth and the life which unfolds on it are like two actors engaged in constant dialogue.

At this point the critic of the random mutation theory would probably say something like this: Too few mutations occur to provide raw material for all the efficient traits which develop in various species. Moreover, the laws of probability suggest that most mutations would have a detrimental effect on the species. Thus species which passed on mutant traits would rapidly decline. They would *not* evolve into higher forms.

But nature found a solution to this problem as well, a solution called "sexuality." Sexuality involes a special kind of relationship between two organisms of the same species. Human beings often view this phenomenon from an exclusively subjective point of view. We perceive sexuality in terms of the contact between two *individuals.* Thus we often ignore its implications for an entire species. The sexual tie developed very early in the course of evolution; biologically speaking, it developed for a good reason.

Until recently even scientists did not know what to make of the phenomenon of sexuality. All of us have such strong personal feelings about sexuality that we cannot really view it objectively. Man has an instinctive tendency to view his world from an egocentric perspective, and the tendency grows more pronounced where sexual matters are concerned. Human subjectivity has colored even the scientific analysis of this field. Thus once we achieve true objectivity regarding sexuality, we are all the more startled by what we see. We feel that we have glimpsed a new horizon. In this case, the "new horizon" lies in the recognition that sexuality more or less represents the "imagination" of nature.

We have observed that certain conflicts in nature unfold on a purely preconscious level. We noted that at the outset of the evolutionary process, nature was confronted with a basic dilemma: how to preserve advantageous traits without sacrificing the capacity for change. The species had to be able to change in order to adapt to a changing environment. Nature had somehow to reconcile tradition with progress. In order to achieve this reconciliation, nature had to confront the same basic choices which human beings confront in trying to conduct their lives.

We may carry this analogy somewhat further. Nature, like man, confronts basic choices. Heredity may be compared to human memory. It constitutes the "memory" of a whole species. Human beings remember everything they experience. In fact, to "gain experience" actually means to store up various incidents in one's memory. The more experiences I have stored up, the more these memories determine my behavior and the future course of my life. We all know that as a person grows older, his personality becomes more sharply defined. He evolves; but at the same time his temperament and attitudes inevitably grow more fixed. A child too has a basic temperament; but he can still grow in many directions. As the child grows up, one possible direction of growth after another must be sacrificed. Life consists in the necessity of choosing only a few out of many possible alternatives.

Precisely the same thing has occurred in the evolution of life itself. When life first began, it took the form of primitive organic molecules. These molecules became the basic building blocks of all future life forms. When these primitive entities first came into being, the "sky was the limit." Any number of different organisms could have been built with these elementary building blocks of nature. Thus far terrestrial life forms have not even begun to exhaust the possibilities inherent in the first molecules. Nor will most of these possibilities be realized in the future. When the first of these potential life forms came into being, nature had just "had her first experience." The "memory" of the species began to store up every detail of each organism's construction. It memorized each nerve fiber and muscle cell. It also stored up functional details: how to split up a sugar molecule to produce energy; how to divide atoms of ionized metal in the cell wall in order to produce the electrical tension which conducts a nerve impulse. But each time one of these details was "memorized," two things occurred. First, the detail became a permanent inherited trait of each species. Second, the unlimited capacity for future growth was somewhat diminished. Any change which might take place in the future would have to build on a foundation which had already been laid; it would have to adapt itself to already existing conditions.

The number of different species continued to increase; their complexity increased as well. But as life developed, it lost its initial flexibility: it could no longer branch out in all directions. The life forms on our planet became more specialized; their individual traits grew more sharply defined. We have no example of extraterrestrial life at our disposal; therefore we have nothing with which to compare our earth species. Because we have no grounds of comparison, we fail to recognize how specialized terrestrial life forms really are. Just as an individual personality matures by narrowing and intensifying its traits, a species develops by retaining what it has already experienced.

Thus heredity functions as a sort of racial memory. This statement may sound like a mere poetic conceit. But a startling recent discovery revealed how literal the analogy is.

For the past ten years scientists have been conducting some fascinating experiments. They wished to discover whether simple learned behavior could be transferred directly from one laboratory animal to another. This learned behavior might consist of simple tricks which an animal had been taught in the laboratory. Could the knowledge in the brain of one animal actually be transmitted to the brain of a second animal simply by means of injections? American scientists in particular have been doing intensive reasearch in this field. Eventually they obtained some positive results. Not surprisingly, these results were greeted with a good deal of skepticism. But now it has been conclusively proven that behavior learned by one trained animal can be physically transmitted to an untrained animal. The learned material is transmitted through brain fluid extracted from one animal and injected into the brain of another.

These fascinating experiments have opened up many new avenues of research. Scientists now have an opportunity to study the actual matter which composes our memories. Specific experiences can be transferred through injections of brain fluid; therefore the substances which compose memories must somehow be contained in the fluid itself. Scientists have discovered that memories can be physically stored up in various substances. Actual experience can be transmitted from one creature to another by merely transferring these substances. The most important of these substances appears to be the complex molecule which biologists call DNA for short. (Its true name is deoxyribonucleic acid.)

Actually biologists had been familiar with DNA for some time before they discovered its memory-storing properties. They first became aware of DNA while exploring an apparently unrelated problem: DNA is the substance which enables nature to store up hereditary traits in the genes. Thus nature seems to have "remembered" an earlier invention and used it again to solve a similar problem. In both cases the basic problem is that of storing up a learned pattern. Nature uses DNA to store up memory traces as well as hereditary traits.

Thus DNA preserves the faculty of memory. Genetic mutations, on the other hand, play the role of inspiration. A human being never becomes totally fixed in his attitudes or behavior. His memory is filled with past experiences; yet he remains subject to "brainstorms"—to the process of involuntary thought. His "involuntary" thought patterns remain partially independent of the attitudes and tendencies ingrained in him throughout his personal history. His thoughts arise "by chance" and are frequently meaningless. But these random or meaningless thoughts perform an important function: They preserve the individual from one-sidedness and offer him opportunities for change. Genetic mutations perform this same role for an entire species. If no mutations occurred, then a species would become rigid and inbred; it would continue to reproduce the same traits over and over. In this way it would lose the capacity to adapt to long-term changes in environmental conditions.

Now we can better understand the decisive role which sexuality plays in the course of evolution. We shall soon see why sexual reproduction must be biologically superior to any other form of reproduction nature could devise. Earlier in our discussion, we noted one basic objection to the random mutation theory. This argument suggested that most random mutations were harmful to the organism; therefore mutations could produce only harmful or deadly traits in a species.

A single genetic mutation might be compared to a "bright idea" in the life of a human being. This analogy can help us to understand how sexuality prevents random mutations from adversely affecting a species. Sexual reproduction enables an organism to give all mutations a "trial run" before actually incorporating them into the species. Thus sexuality plays the same role in the evolution of a species as imagination plays in the life of an individual.

We actually act on only a few of the ideas that come into our heads. One does not need to be a psychoanalyst to realize how many unacceptable thoughts end up being repressed by our mind or our conscience. We are constantly filled with thoughts which remain below the threshold of consciousness. Our species, *homo sapiens*, is basically distinguished by one trait—the ability to think of something without immediately acting upon it. This ability has made man lord over the earth. An animal adapts to its environment by means of instinct. Instinct represents a form of innate experience. At times an animal finds itself in an unfamiliar situation; now its repertoire of instincts cannot tell it how to behave. The animal cannot react with any of its "prefabricated" behavior patterns. In this case, it can only do one thing: It must fall back on the risky "trial and error" method. It must simply try out every impulse in order to find which experiment works. Chances are that this procedure will prove painful, perhaps even fatal. An animal's first experiment may well turn out to be its last.

Man's superiority to other creatures lies in his ability to anticipate events and their consequences. He "plays things through" in his imagination before committing himself to carrying them out. Man's capacity to develop an "internal model of the outside world" places him in a unique position. Without exposing himself to any danger, without committing himself to any course of action, he can test out various solutions to a problem. He has the freedom of a child building and tearing down buildings made of blocks. After evaluating the alternatives, he can decide on a particular course of action. This sort of mental "dress rehearsal" occurs in day-to-day life situations; but it also takes place during the formation of a new scientific theory or a work of art.

This special ability which characterizes our species does appear to have its limits. Man today seems to be creating environmental problems—the population explosion or the threat of nuclear annihilation—which lie beyond his power to control. The human imagination may prove incapable of "playing through" the consequences of such problems. If we cannot learn from a dress rehearsal, we may have to learn from the real thing. In any case, we need not meditate further on this dire possibility. We are concerned here with the faculty of imagination itself.

Recent research into the laws governing evolution has revealed that human imagination does not represent a wholly unique phenomenon. Nature has used the principle of the "child's blocks" before. That is, in the past nature devised a method for testing new possibilities before they become reality. We have noted that sexual reproduction performs the same function for the species as imagination performs in the life of the individual. A one-celled organism which reproduces by dividing is an extremely vulnerable organism. Any mutation which occurs will at once manifest its effects; and most mutations are in fact harmful. Thus only a tiny percentage of these creatures with mutant genes will even survive a mutation; fewer still will derive any advantage from it. Most such primitive species reproduce at an incredible rate; some bacteria pass through three generations in a single hour!

This swift rate of reproduction helps to balance out their high mortality rate. The balance between birth and death enables the species to survive, but only at the cost of the individual organism. Moreover, these primitive species suffer one added disadvantage from their asexual mode of reproduction: They evolve very slowly. Their ability to adapt more advantageously to their environment is regulated wholly by chance.

Species which rely on sexual reproduction confront a more hopeful future. All the higher animals, most plants, and even many one-celled organisms reproduce by this method. This fact in itself proves that sexual reproduction must offer some distinct advantage over other reproductive methods. Its basic advantage resides in the fact that all creatures which reproduce sexually carry *two* identical sets of chromosomes in their cells. Cells of this sort are called "diploid." The various genes are strung along the chromosomes like beads. In sexually reproductive organisms, each of these gene-filled chromosomes occurs twice. Thus a mutation in a single gene does not have any immediate effect; the corresponding gene in the "twin" chromosome has not been similarly affected. The mutant gene is more or less "covered up" by the normal twin. But sexual reproduction equips living organisms with a number of other distinct advantages.

When an organism reproduces by simply dividing, each of the two new individuals develops from one half of the parent cell. Sexual reproduction, on the other hand, arises from the combination of two "half cells" taken from two individuals of the same species. The sexual organs of both parents contain cells known as germ cells. Each of these cells contains only *one* set of chromosomes. Thus when the germ cells of the two parents unite, the new individual will acquire the normal "diploid," or double, chromosome set. As long as a germ cell remains "haploid," each of its chromosomes appears only once. This temporary haploid condition helps to protect sexually reproductive species against the threat posed by harmful mutations. We have noted that if one of two chromosomes is a mutant, its twin will tend to "cover" it, that is, its twin will prevent it from affecting the organism. During the haploid stage, no twin exists to "cover" a mutant gene. This haploid stage does not occur in the organism as a whole, but only in the life of an individual germ cell. Here in the germ cell, nothing hinders the development of a mutation. This arrangement enables any new mutant gene inside the germ cell to "try out" its effects on the organism.

The germ cell functions as a sort of test tube. The mutation can be tested here without endangering either the organism or its offspring. The "test" lies in finding out whether the mutation will prove compatible with the complex metabolic processes of the germ cell. Most mutations would contribute nothing useful to the organism; many are actually harmful. After all, mutations result from "errors" in the reproduction of already-existing genes. The germ cell screens and eliminates all these unsuccessful mutations in the most efficient way imaginable; if the test results are "negative," the cell itself simply dies! Thus the mutation contained in the dead cell has been quietly discarded. Only the single germ cell suffers any ill effects; the organism itself remains unaware of the whole process.

No doubt each day countless mutations are eliminated from our systems. At the same time, they are being eliminated from further competition in the evolutionary struggle itself. Every new hereditary trait must satisfy one basic requirement: compatibility with the life processes of an organic cell. Unless a mutation can satisfy this requirement, it cannot cross the barrier erected by nature. To be sure, this seems a very modest requirement. Many mutant genes quite capable of passing this first test might still exercise some harmful effect on an organism. Therefore, any new, "progressive" hereditary trait must also prove its compatibility with all the "traditional" traits of a species.

So far we have only partially refuted the critic of the random mutation theory. At this point he may still be asking, what happens to mutant genes which pass the first test—compatibility with a single living cell—but which might harm the organism as a whole? After all, any higher organism is composed of many hereditary traits. What happens if the new gene comes into conflict with the older established genes?

What happens is this: New genes are stored up in the "recessive gene pool" of the species. Over the course of many generations, these genes will be used over and over in various combinations involving other mutant genes.

During sexual reproduction a germ cell can produce a new individual only if it can unite with the germ cell of the other parent. Both parent cells contain exactly the same set of chromosomes. From the moment the two cells unite, they pass into a diploid stage. Two genes now determine each trait. It hardly ever happens that a germ cell containing a mutant gene will unite with a germ cell containing precisely the same mutation of precisely the same gene. Even a simple bacterium contains 10,000 different genes. Higher organisms often contain millions. Thus it almost never happens that two identical mutations combine at the moment of fertilization.

There is one exception to this rule: the case of persistent inbreeding. The smaller the circle of available sex partners, the smaller the number of genes in the gene pool. The genes in the gene pool continually enter into new combinations. If there are only a few available genes in the pool, then chances increase that sooner or later two germ cells containing the same mutant gene will join to form a new individual. In this case, the individual in question will have no protection against the mutant trait. More than likely, this mutant trait will exercise a negative effect on the organism. Like human beings, most higher animals quickly developed taboos and special behavior patterns to circumvent this problem. These taboos prohibit the selection of sexual partners from within the organism's own immediate group.

In our school days, we all learned something about recessive genes. If a mutant gene has no effect on the organism it inhabits, biologists refer to it as "recessive." The effect of the mutant gene is inhibited by the normal gene of its twin chromosome. Recessive traits can be "invisibly" stored up in an organism, thus providing the raw material for new genetic combinations in future generations.

Until recently biologists could not have explained the real advantage of a reproductive method involving the use of recessive genes. After all, a mutant gene can aid a species to evolve only if the gene succeeds in producing some actual change in the organism. Only then can the mutant organism confront its environment and be exposed to the process of natural selection. At this point the opponent of the random mutation theory would probably say that the presence of recessive genes simply postpones the inevitable: Sooner or later the recessive gene will produce a mutant organism. One day a confrontation must occur between this organism and its environment. What is the use of simply postponing the day when the genes will manifest their harmful effects? Biologists now know how to respond to this argument.

Amazing as it may sound, selective mechanisms exist within the recessive gene pool itself. Over the course of many generations, the recessive genes form different combinations. Throughout this process, they are constantly being tested. They are "selected" in the same way that the human imagination selects one among many alternatives. The selection process does not unfold in the real world, but in the microscopic realm of the cell nucleus; the selection occurs invisibly and never affects the organism directly. In fact, we have found another instance of the "building block model": the ability to "play with" various models of reality without incurring any real consequences.

How does this internal process of selection take place? Certain genes in the cell nucleus test out various combinations of new and "old" genes. These so-called "regulator genes" must ensure the compatibility of each mutant gene with the genes which have already proved their worth. They must test out all these combinations inside the cell nucleus and monitor the results. This procedure affords an entire species the same advantage as imagination affords a human being. The "regulator genes" can select those genetic combinations which seem best suited to the already-existing organism. These selected mutations will offer the species the best possible chance of adapting more successfully to its environment.

There are also "reparative" or corrective genes which can reverse a harmful mutation some time after it has actually taken effect. These genes simply restore the mutant gene to its original premutant state. Probably they alter the mutant gene to match the normal gene of its twin chromosome. Other "regulator genes" check up on each new organism; they make sure that all its new mutant traits are compatible and that none of them are producing mutually contradictory effects.

Biologists do not really know how the regulator genes achieve their effects. Nor is it clear why these genes "select" certain mutations rather than others. Scientists have been doing research in this area for only a short time. They have not yet succeeded in unlocking the complex microscopic world of the cell nucleus. But biologists believe that all creatures equipped with a double set of chromosomes possess a sort of "internal model of the external world." This "internal model" seems to be located somewhere in the cell nucleus, where it plays an invisible but essential role in the process of selection. A model of this sort greatly increases the chances that a species will continue to evolve. But only sexually reproductive species possess this special "imaginative" faculty.

In school we may have learned a little about the science of genetics. But recent research has revolutionized our former image of the gene. Geneticists estimate that only about five percent of the genes in a single cell function as "structural genes." These are the genes which actually determine the hereditary traits of an individual organism. This means that ninety-five percent of all genes are probably regulator genes. Since every higher animal has a total of about a million genes, 950,000 of these must be involved in the processes of selection. This amazing figure reveals how important the internal selection process must be to the biological evolution of a species.

Thus the cell nucleus acts as the "playroom" of evolution; and sexuality acts as its imagination. The actual "playthings" with which this imagination plays are the mutations themselves. These mutant genes are like the tiny colored tiles of a mosaic. At first the tiles simply form a confused and multicolored heap. Each tile is meaningless by itself. But in time all the tiles can unite to form the complex pattern of a living being. Not every tile will fit into every position; some of them will not fit anywhere. The mason must not use too many of them, for this would spoil his design; rushing his work would ruin the pattern too. Evolution, the master builder, keeps on creating new and more complex patterns. In order to carry out his designs, the master builder must keep on hand a ready supply of tiles in various shapes and colors—the various mutations in the gene pool. Only then can the evolutionary "imagination" freely experiment with various possibilities.

To some degree, the tempo of evolution depends on the number of available "tiles," or mutations. If very few mutations occur, then a species ceases to develop. This species has adapted to a specific environment; it will survive only so long as the environment remains unchanged. It no longer has the "flexibility" which enabled it to adapt to its surroundings when it first evolved. A species may be regarded as "successful" when it has adapted well to one specific environment. The more specialized the adaptation, the more control a species has over its surroundings. But such a "successful" species becomes all the more vulnerable if conditions around it suddenly begin to change. It becomes the victim of its own success, of its own ideal adaptation. Any prolonged disruption in the climate, a change in the vegetation which supplies food, the arrival of a new competitor for a limited food supply—these and many other seemingly minor changes can swiftly lead to disaster. A certain form of adaptation no longer "fits." A species which confronts such changes in its biological milieu must be able to respond with a new genetic adaptation. If it cannot do so, it simply becomes extinct. This seems to have been the fate which befell the dinosaurs—one of the most "successful" species which ever existed on earth.

Thus if too few mutations occur, a species virtually ceases to evolve. But too many mutations can also halt the course of evolution. A sudden deluge of new traits will upset the balance of the older traits, thus disrupting the entire organism. Recently scientists have experimented with the effects of radiation on hereditary traits. For example, radioactive or X rays alter the molecular structure of the genes, thereby producing a flood of new mutations. When cells are subjected to very intense radiation, the number of mutations rises sharply. Soon the protective and selective mechanisms in the cell nucleus are overwhelmed by the mutant genes. Increasing numbers of harmful mutations appear in succeeding generations: deformed bodies, defective organs, metabolic disturbances, and various other anomalies. A species overwhelmed by this sort of disaster must inevitably be destroyed. Presumably human survivors of a worldwide nuclear war would suffer the same fate. Radioactive isotopes would poison the atmosphere for centuries to come.

Evolution can only take place somewhere between the two extremes, neither too many mutations nor too few. Life depends on the maintenance of many complex processes. Living creatures must be able to change gradually, without endangering that complex structure which has already been achieved. The history of life has unfolded between these two limits, stability and change. The tempo of evolution has constantly varied. When few mutations occur, things change very slowly. It is easy to see why this should be so. Our "master builder" must wait a long time before enough "tiles" are produced to provide just the one he needs to complete a particular pattern. If more mutations occur, the pace of evolution accelerates once more. In this case, the "master builder" probably has just the right tile on hand to use when he needs it. The tempo of evolution may continue to increase until it reaches the critical point. Beyond this point, the abundance of mutations begins to destroy the order which has already been achieved.

Days of the Dinosaurs

WE HAVE briefly discussed some of the basic principles underlying the evolutionary process. Now we can more readily understand how the periodic collapse of the geomagnetic field affected life on earth. The earth repeatedly lost its magnetosphere for periods of at least one thousand years. Throughout these periods, our planet lay exposed to the solar wind.

In a sense the earth was not really "exposed" at all. The particles of the solar wind did not actually succeed in reaching the earth's surface. Our atmosphere "captured" them, preventing them from continuing their journey. On the other hand, these particles did bombard the upper atmosphere with far more than their usual intensity, creating far more than the usual quota of "secondary radiation." This by-product took the form of radioactive C^{14} and other radioactive isotopes.

The renowned American geneticist C. H. Waddington offers some expert testimony on this point. During the long scientific debate over the biological effects of dipolar reversals of the magnetic field, Waddington sided with the second group. This group maintained that the periodic loss of the magnetosphere had not affected living creatures on earth. Waddington quite correctly assumed that the earth's atmosphere would have prevented the solar wind from actually reaching the earth's surface. But even Waddington conceded that during the dipolar reversals, the solar wind must have created at least twice the usual number of radioactive isotopes in the earth's atmosphere. Moreover, he believed that solar flares, which erupt every few months from the sun's surface, periodically intensified the solar wind bombardment. Waddington supposed that solar flares must increase the quantities of radioactive isotopes in the earth's atmosphere to an even higher level than did the periodic loss of the magnetosphere. The group of radioactive isotopes included not only C^{14}, but also beryllium with its half life of 2,500,000 years, and various other elements.

When the quantity of radioactive isotopes in our atmosphere increases, the motor of evolution immediately picks up speed. The speed of this motor normally depends on the number of random mutations being produced within a species. Mutations represent "spontaneous" genetic leaps, that is, they do not follow any sort of pattern and do not develop to fulfill any specific function. When scientists refer to mutations as "spontaneous," they do not mean to imply that these mutations have no immediate *cause*. We do not yet know how many factors contribute to the production of mutations. We have already mentioned the experiments investigating the influence of intense radiation on hereditary traits. These experiments and other sorts of data have revealed at least one of the factors involved in the production of mutations—radiation. The number of mutations which occur in a given species depends on the intensity of the radiation in their environment.

The radiation needed to produce mutations has existed on the earth from the very beginning. We have already mentioned the phenomenon of "background radiation." This radiation appears to be composed of radiation arising from several different sources. First, it is fed by radiation emanating from radioactive elements in the earth's crust. Furthermore, it consists of the cosmic radiation which travels to earth from somewhere out in the Milky Way, some of which succeeds in penetrating the protective sphere which the solar wind has erected around our entire solar system. Finally, background radiation is fed by the radioactive isotopes created by the solar wind as its particles bombard the upper atmosphere.

Background radiation may be weak, but it is constant, and strong enough to be recorded on our instruments. Radiation plays a basic role in our environment and has profoundly influenced all living organisms on earth. We have not yet discovered other factors which may contribute to the development of mutations. If mutations serve as the "motor" of evolution, then background radiation acts as the throttle which determines the motor's speed. That is, radiation controls the tempo of evolution.

It should now be clear why the reversals of the magnetic poles must have profoundly influenced the history of life on earth. Each time the poles were reversed, the larger quantities of radioactive isotopes in the earth's atmosphere markedly increased the intensity of the background radiation for at least a thousand years. Thus the supply of mutations suddenly increased too. Evolution received a shot in the arm. Each time the earth lost its magnetosphere, a new act was written in the drama of evolving life. The acceleration of the evolutionary process threatened all the dominant life forms of that time; every traditional trait was suddenly called into question. At the same time, a few new species just beginning to evolve were dealt a winning hand.

The accelerated tempo of evolution simultaneously affects species all over the earth. These species confront a bizarre situation which we can best understand in terms of an analogy. Suppose that a man is trying to show some colored slides to his friends when suddenly a child comes into the room and begins playing with the slide projector. The child knows nothing about how the projector works and twists one of the knobs back and forth without paying any attention to the results, so that the slides on the screen keep going out of focus. The man showing the slides had adjusted them so that they were in very sharp focus. Thus any change the child makes seems a change for the worse. But suppose that certain slides have not yet been brought into sharp focus. The audience will notice that some of the slides are accidentally being *improved* by this senseless game. In fact, a few have suddenly come into perfect focus! Eventually the audience will observe that the chances of sharpening the focus of a slide depend on how sharply it was focussed to begin with; the more blurred an image seemed at first, the better its chances of growing sharper through the random twisting of a knob. If an image is in perfect focus at the start, it has no chance at all of retaining its clarity.

Let us play another mental game analogous to the first. Suppose that a cosmic observer is sitting somewhere out in space so that he can take in the earth and all its creatures at a single glance. He witnesses all the events which take place during and after a reversal of the magnetic poles. He sees the giant sphere of the magnetic shield begin to shrivel and disappear. Shortly thereafter, all terrestrial life forms suddenly increase their production of mutations; this increased production has been triggered by the increased intensity of background radiation. The background radiation in turn has been intensified by the radical increase in the number of radioactive isotopes in the earth's atmosphere. The increase in the number of mutations accelerates the tempo of evolution. *Apparently* every species has been granted an equal chance to speed up its own development.

Very soon our hypothetical cosmic observer would discover that all the species have *not* been given an equal chance. Those species which have been highly successful up to now have not continued to develop; nor are their "inferior" competitors losing ground. On the contrary, the observer notes that a number of hitherto insignificant organisms are suddenly beginning to thrive, developing into completely new species and producing a variety of new life forms which are successfully adapting to the most widely varied environments. On the other hand, those species which have been dominant up to now have fallen into a decline. In most cases they are vanishing altogether from the stage.

No wonder that, when the hitherto unchallenged lords of the earth are exposed to the increased number of mutant traits, they do not develop further. After all, they were the dominant species precisely because they had already achieved the optimum adjustment to their own environment. How could they possibly derive any real benefit from a deluge of new mutations? An image which is in perfect focus can only grow blurred if subjected to any change.

Our cosmic observer would shake his head sadly at what he saw next. The dominant species are suddenly turning into caricatures of their former selves: They develop deformed bodies and grow to unprecedented size. Various anomalies disrupt the harmonious structure these creatures had achieved in the course of their development. To make the disaster complete, while the former masters grow weaker, some of their competitors are suddenly growing stronger. A multitude of new and vital life forms now develops. These new life forms are the descendants of a few insignificant organisms which up to now have functioned as the "wallflowers of evolution." The new forms have benefited from a long chain of cosmic events. These formerly awkward and insignificant beings have suddenly come into focus—just like the blurred images cast by the slide projector. In their case, any change at all might well be a change for the better.

To be sure, only a few fortunate species actually benefit from the altered conditions—just as only a few of the blurred slides would actually come into better focus. Most species fail to profit from their hour of destiny. The lucky ones set off fireworks of new species and new modes of adaptation. A new act in the history of life has begun.

Most of us looking at this scene would recall the fate of the dinosaurs when they were supplanted by the first mammals. For years no one was able to explain this disaster. Many people know the dinosaurs by their nickname, "giant lizards," but not all the dinosaurs were giant. Their family produced the largest creatures which have ever walked the earth; but it also produced a number of very small creatures. More important than their size is the fact that the dinosaur family developed so many subspecies. Dinosaurs lived on the land; but fishlike dinosaurs also inhabited the seas. Dinosaurs flew through the air on webbed wings like those of the bats which developed much later; their flying technique also resembled that of bats. Some dinosaurs were savage carnivores; but most were herbivorous. A biologist would say that the dinosaurs had occupied the most attractive and ecologically advantageous areas of the earth's surface, leaving the remaining space to all the other creatures. Thus they ruled the earth unchallenged for the inconceivable span of more than 30,000,000 years. It must have seemed that their reign would never come to an end.

During these 30,000,000 years, the dinosaurs must have lived through several dipolar reversals of the earth's magnetic field, and at first they probably profited from the periodic increase in the level of background radiation. No known major disaster occurred to which we might attribute their sudden startling decline. But something clearly happened around 200,000,000 years ago which gravely upset their balance. Perhaps they were exposed to another temporary acceleration of the evolutionary process. It might be that by this time they had already realized all the inherent potential of their kind; perhaps all their variant species had achieved their optimum degree of adaptation. If true, this fact may help to explain why the dinosaurs were so rapidly defeated by their rising competitors. A new type of organism was entering the stage, a little creature scarcely the size of a shrew. This rather absurd beast had just "discovered" the phenomenon of warm blood; he was a forebear of present-day mammalian life.

Unfortunately we have not yet succeeded in proving this hypothesis. But whatever may have happened in the age of the dinosaurs, we can at least be sure of one thing: The dipolar reversals of the magnetic field decisively influenced the history of life on our planet. Every few hundred thousand or million years, these reversals intervened in the course of evolution. Here we have another example of that vast cosmic network which unites our everyday world with forces reigning outside the earth's atmosphere. If this network did not exist, then neither the earth nor the creatures which live here would ever have come into being. And if the forces beyond our atmosphere should ever cease to reach out to us across space, neither the earth nor its creatures could survive.

How strange to realize that this ever-changing relationship between our little earth and its sheltering magnetosphere mirrors the events inside a cell nucleus! In both cases, the basic conflict is the same: how to reconcile the contradictory demands of stability and progress. In the one case, the dimensions of the conflict are microscopic, in the other, astronomical. During "normal" intervals, the magnetic shield ensures the moderate tempo of evolution which benefits already-existing, fully developed forms of life. During these periods, the earth seems to stop and draw breath. These "breaths" last a few hundred thousand or even a few million years. But then the mutability of the magnetic shield reasserts itself once more: a mighty force intervenes in the course of evolution, accelerating evolutionary development. The accelerated tempo opens the door to new life forms—at the same time shutting it on the enduring forms which have already reached their zenith. This is a "cosmic" relationship in every sense of the word; it reveals a vast hidden order in both microcosm and macrocosm. The behavior of the magnetosphere is related to the processes unfolding in the nucleus of a living cell. And both realms—

the microscopic and the astronomical—contribute toward the same end, the creation and maintenance of life.

When scientists first discovered the phenomenon of the dipolar reversal of the geomagnetic field, they disagreed as to its effects. From the very beginning, some scientists viewed the situation more or less in the way we have just described. That is, they believed that the reversals of the magnetic poles had profoundly affected life on earth. Other scientists opposed this view. Recent discoveries have finally settled the debate. Scientists have found evidence directly linking the extinction of certain species with periods when the earth lacked its magnetic shield.

We shall examine one especially noteworthy example of these recent discoveries, a contribution made in 1967 by the American oceanographers Billy Glass and Bruce Heezen. The two men gathered conclusive evidence that the reversals of the magnetic poles had influenced the course of evolution. Moreover, they found clues to the possible *cause* of the dipolar reversals. Once again it appears that a powerful force has been reaching out to us from the depths of space. Scientists now believe that the periodic loss of the magnetic shield was produced by violent cosmic collisions between the earth and gigantic meteors weighing hundreds of millions of tons! As they struck the earth, the meteors disrupted the even rotation of the earth's liquid core, which may serve as a dynamo to produce the magnetic shield. The impact of the collision presumably created areas of turbulence within the liquid portions of the earth, so that the liquid ceased to flow in a single direction. When the "dynamo" ceased to produce the magnetic field, our protective shield fell away. Then gradually the influence of the moon probably started up the dynamo inside the earth again, thus restoring the magnetic field.

But we must take our time in telling the tale of this astounding discovery. To begin with, we might ask why this discovery should have been made by two oceanographers, rather than by geologists, geophysicists, or astronomers. The explanation lies in the fact that scientists had just invented a new technique; they had found a way to investigate paleomagnetism even in the depths of the sea.

In 1966 the American geologist John Foster conquered the technical problems involved in the investigation of deep-sea fossil magnetism. Foster designed an instrument which enabled researchers to obtain drill samples of undersea volcanic rock from various layers of the ocean floor, hundreds or thousands of feet beneath the surface. On the surface, the drill samples were tested to determine their paleomagnetic orientation. Clearly it was not an easy task to overcome the technical problems connected with undersea drilling. Drilling at such depths requires the use of extremely long drilling rods. Moreover, Foster had to devise a great deal of additional equipment to meet the special demands of drilling for magnetic rock. Once the test samples were safely on board ship, researchers needed various complicated instruments to determine their magnetic orientation. Their special instruments enabled scientists to reconstruct the original polarity possessed by a given rock sample while it still formed part of the ocean floor. Unless scientists could accurately reconstruct the original polarity of the various rock layers, they would be unable to compare them.

Before Foster's invention, scientists had been able to study fossilized magnetism only on the exposed surface of the earth—on its continental land masses. They had long wanted to investigate this phenomenon on the ocean floor, particularly because they hoped the magnetic record of the earth's past history would be better preserved beneath the sea than on the land. On land, wind and water had been wearing away the layers of volcanic rock for millions of years; at times bits of rock were scattered far from their source. But the rock layers under the sea had been protected by the ocean waters.

In 1966 the technical problems of deep-sea drilling were at last overcome. At once scientists began their investigations. Within the first few days, their faith was rewarded; the ocean floor revealed something they could not have found on land. Deep-sea volcanic eruptions differ from those on the continents. Hundreds or thousands of feet below the water's surface there are still active volcanoes. But the lava which flows from them is cooled by the sea water the moment it emerges from the crater; it hardens almost at once. This abrupt hardening of the lava sometimes produces quite remarkable formations, the most famous of which are the volcanically active trenches which scientists have discovered in the floors of all the major oceans. Apparently lava has been flowing from these trenches for millions of years. The lava forms broad carpets which spread in an even plane across the ocean floor on both sides of the trench. On the open surface of the earth, the lava layers from successive eruptions lie one on top of another. Beneath the sea, the various lava layers spread out *beside* each other. The oldest lava emissions lie farthest from the trench; the most recent emissions harden right beside it. In other words, these trench areas form a visible calendar of successive geological epochs.

When scientists began to study the paleomagnetic orientation of these undersea lava carpets, they quickly discovered records of the various dipolar reversals of the geomagnetic field. The evidence from the depths of the sea verified the evidence of the basalt rocks on the earth's surface; both areas recorded the same intervals separating the reversals of the magnetic poles. But unlike the rocks on land, the undersea lava carpets lay next to each other rather than on top of each other. We can draw a geological diagram of one of these oceanic regions, depicting the various zones of lava, each with its own magnetic orientation. The result is a sort of striped, zebra pattern.

Apart from the fact that undersea volcanic rocks were better preserved than those on land, a second factor also attracted scientists to the paleomagnetic investigation of the ocean floor. They had reason to believe that a stream of tiny metallic particles was constantly raining down on the earth from outer space. The particles were small enough to land "softly" in the upper layers of the earth's atmosphere. Thus they did not burn up in the atmos-

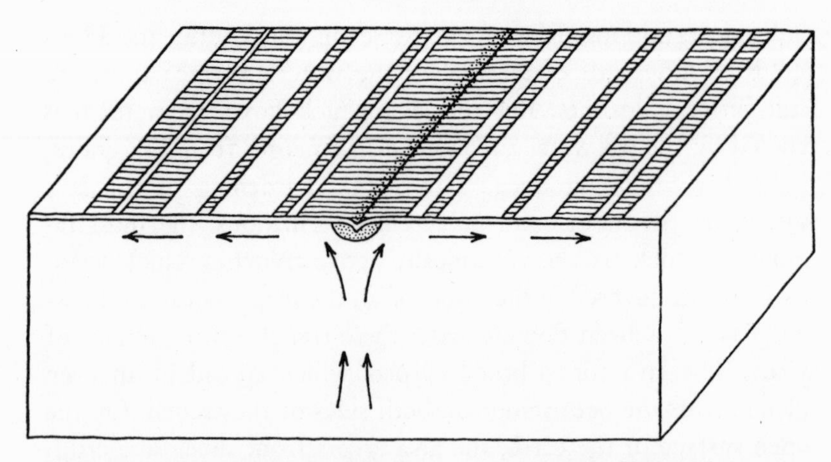

The earth's magnetic North Pole has not always coincided with its geographical North Pole. During the past 76,000,000 years, the earth's magnetic field has reversed its poles at least 170 times. Certain regions of the ocean floor have been uninterruptedly emitting lava for millions of years. These areas record the history of the repeated dipolar reversals of the magnetic field. Scientists can "read" the lava record like a calendar.

phere as large meteors almost invariably do. Scientists believed that some of these particles represented fragments of formerly larger meteors. Even though the particles were made of metal, they were small enough to drift downward just as dust particles do. Thus, unlike the larger meteor fragments, they eventually arrived unharmed on the earth's surface.

So small were these "heavenly" dust particles that no one could actually see them; but scientists were convinced of their existence. There was no lower limit to the size of meteorites. Thus meteor dust ought logically to exist. Moreover, the number of meteors observed on earth was always inversely related to their size; the smaller the meteors, the more frequently they appeared. Large meteors were extremely rare. Thus everything indicated that meteor dust must be extremely plentiful. But scientists could not verify its existence on the land, since the bewildering quantities of minerals and metals on the continents made it impossible to identify those particles which had come from somewhere out in space. Geologists believed that the soil samples they had so often observed under their microscopes must have contained some particles of extraterrestrial dust, but no one could tell just by looking whether dust came from this world or another.

A hundred years ago, a Swedish scholar was thinking about the problem of distinguishing meteor fragments from the ordinary dust of the earth. He undertook an experiment which probably made his neighbors think he was a bit peculiar. One day he began to shovel huge masses of freshly fallen snow, and spent the next few days melting it in a huge cauldron. This must have seemed rather eccentric behavior on the part of a supposedly serious scholar; but there was method in his madness. When Adolf Erik, Baron von Nordenskjöld, "geognost and polar explorer," had boiled the snow to a fine-grained blackish sediment, he began to sift through this material with a magnet and to study the magnetic samples under his microscope. He soon found exactly what he had hoped to find: tiny metallic dust particles with magnetic properties.

Von Nordenskjöld immediately gave a lecture, informing his audience that the virgin snow had contained metallic dust from somewhere far out in the cosmos. Apparently the fine dust had slowly filtered down through the atmosphere, where it served as a core for crystallizing snowflakes, which then bore the dust to earth. The lecture audience in Stockholm politely applauded this announcement, but they remained unconvinced. We now know that Von Nordenskjöld's assumption was quite correct. Unfortunately he was unable to prove it.

Scientists before Von Nordenskjöld had been aware of the same phenomenon in another form. In 1845 the great German naturalist Alexander von Humboldt published his book *The Cosmos,* containing a summary of all the scientific knowledge of his time. In this work, Von Humboldt mentions a fact that scientists had observed regarding the familiar cirrus clouds which floated high up in the atmosphere: The cloud streaks often lined up parallel to the poles of the earth's magnetic field. Von Humboldt merely *mentions* this fact; he is unable to explain it. We now know that this behavior of cirrus clouds reflects the same phenomenon as Von Nordenskjöld's snow experiment, which took place twenty-five years later. Like the snowflakes, cirrus clouds consist of tiny drops of water which have crystallized around grains of meteor dust in the upper atmosphere. This dust has magnetic properties; therefore the long strips of cloud line up parallel to the earth's magnetic poles.

We owe our present knowledge of this phenomenon to an experiment conducted in 1962. On August 11 of that year, scientists finally received a definitive answer to their questions about meteor dust. The method they used typifies the directness and efficiency of twentieth-century technology. They simply fired a rocket directly into a cloud they suspected might contain meteor particles. The nose cone of the rocket was fitted with a special instrument designed to capture microscopic dust particles. After the parachute landing of the nose cone, scientists examined the laminated plates designed to pick up the dust; they found large quantities of metallic particles. This time no one could doubt that the dust came from outside our atmosphere, since scientists had taken care to prevent earth dust from infiltrating the nose cone during the rocket launch or the parachute landing. Thus they knew that the dust could have come only from the cloud itself. The tiny particles, only a few hundred thousandths of an inch in diameter, were subjected to intensive chemical microanalysis. This analysis revealed that they were composed of nickel, iron, cobalt, and copper, the characteristic composition of an ordinary iron meteorite.

Scientists had at last succeeded in establishing the extraterrestrial origin of magnetic dust in the earth's atmosphere. But oceanographers had known for some time that magnetic dust also appeared in the mudlike sediments covering the ocean floor. In some regions these sediments attain a thickness of several hundred yards. Unlike the rock layers on the surface of the continents, the undersea sediments are relatively free of other mineral substances, so that the magnetic dust in these layers can be easily identified. The paleomagnetic experiments conducted on these deep-sea particles produced a new method of testing all previous paleomagnetic data—a method based on a completely different principle.

Up to this point geophysicists had investigated the magnetic orientation of iron-containing minerals in volcanic rock. But meteor dust differs markedly from this magnetized volcanic rock. When the dust arrives in the earth's atmosphere, it is already endowed with magnetic properties. This microscopically fine dust wafts so gently toward the earth that it never heats up to 770° Celsius, the critical temperature at which iron begins to lose its magnetic properties. The tiny dust particles whirl and drift through the air as they fall, being continually caught up in various air currents. Thus they are given no opportunity to line up along the earth's magnetic poles. But as soon as the particles reach the water, everything grows very still. Even the currents which still flow at this level do not divert the particles from their descent. The little metallic chips sink slowly through hundreds or thousands of yards of sea water until they come to rest on the ocean floor. They are free from all outside forces—except one. During this last phase of their descent, they are attracted by the earth's magnetic field. The particles themselves are already magnetized; thus they begin slowly but surely to quiver in a north-to-south direction. In the end, the muddy sea floor is covered by myriads of microscopic compass needles, each of them pointing straight to the earth's magnetic North Pole.

For hundreds of millions of years, an uninterrupted stream of cosmic meteor particles has been settling into the sea. The concentration of metallic dust in marine sediments indicates that each day several thousand tons of meteor dust fall onto the earth's surface. (Von Nordenskjöld's snow experiment may seem primitive to us now; but on the basis of that experiment he arrived at almost this same estimate—one hundred years ago!) The cosmic compass needles have been distributing themselves quite evenly, one layer on top of the next, from the earth's beginning to the present day. The sediments grow very slowly: no more than a few inches or fractions of inches every thousand years. This may sound like a mere drop in the bucket. But if we convert this growth rate into a handy geological unit like one million years, we will see that during this period the sediment will have increased by some five to fifty-five yards. The chronicle of one million years is concentrated into this narrow space.

As soon as the technical problems had been solved, scientists began to retest the paleomagnetic data they had obtained from volcanic rocks, using the cosmic compass needles to verify their earlier findings. To their great joy, the results obtained through this second method completely substantiated the paleomagnetic record of volcanic rocks. Both methods revealed the same number of dipolar reversals of the geomagnetic field; both recorded the same time intervals separating the various reversals. The more deeply scientists penetrated into the past, the more of these reversals they discovered. With respect to dipolar reversals, the marine dust supplied no new data; it simply verified what was already known.

But the cosmic meteor dust did reveal some radically new information. Thus far geophysicists and paleontologists had studied only the basalt rock produced by volcanic eruptions. Lava stone, as we all know, is inanimate matter. Marine sediments, on the other hand, consist largely of organic matter, being principally composed of the remains of countless living organisms that once swam in the waters above the sediment layers. Over the course of thousands and millions of years, the organisms died and sank to the ocean floor. A drill sample from one of these layers contains meteor dust revealing the dates of various dipolar reversals; but this sample will also contain the remains of living creatures which inhabited the ocean at the time of each reversal of the magnetic poles.

After Foster's invention made it possible to obtain deep-sea drill samples, oceanographers immediately applied the technique to the study of prehistoric marine life. For years their less fortunate colleagues the paleontologists had vainly been trying to settle the debate over the biological effects of the reversals of the magnetic poles. In this debate, the paleontologists had been limited to theory and discussion. But oceanographers now saw a chance to prove all the theories one way or the other. All they had to do was to locate a sediment recording the occurrence of a dipolar reversal of the magnetic field, and then to compare the animal remains above and below this sediment layer. Beneath this level they would find an *older* sediment layer, containing organisms which inhabited the ocean *before* the reversal of the magnetic poles. Above this level, the sediment would contain life forms which existed *after* the reversal. Since scientists wished to know how the reversal of the poles had affected living organisms, oceanographers had only to compare the animal populations of the two periods. If there were no notable difference between these two populations, then the dipolar reversals had clearly had little effect on living organisms. On the other hand, marked discrepancies between the two populations would argue strongly in favor of the "disaster party," that group of scientists who maintained that the periodic loss of the magnetic shield had brought about the extinction of whole species.

In reality all of this was easier said than done. Animal remains which have been accumulating for a million years to form a sediment some fifty-five yards thick have long since caked together: it is difficult to tell any of them apart. Most of them decayed long ago; that is, they have been resolved into their separate components. But oceanographers knew of one animal family which might still be identifiable even among the caked sediments miles beneath the surface of the sea. They hoped to find traces of the radiolarians (see Illustrations 21a and b), microscopic one-celled organisms noted for their hard siliceous skeletons. These skeletons are not mere shells like those formed by crabs or snails to protect the abdomen. Long ago, the radiolarians had attracted the attention of biologists for purely esthetic reasons; their protective skeletons assume a variety of remarkably delicate forms. In the past, scientists (among them the famous naturalist Ernst Haeckel) never tired of working with these uniquely beautiful creatures, which are invisible to the naked eye. Illustrations 21a and b reveal the esthetic fascination of these microscopic beings whose beauty normally remains hidden from all eyes.

The beauty of the radiolarians also had its practical consequences. Despite their variety, radiolarian skeletons fall into several basic classes: spheres, crested formations, cylindrical shapes, stars, spears, and a number of others. These characteristic shapes provide a basis for classifying the radiolarians into various species and subspecies. Their siliceous armor does not decay. Thus, under a microscope, scientists can count the various skeleton types and determine their relative abundance in a given epoch. Like all one-celled organisms, the radiolarians are a very ancient species, so that scientists are able to establish the distribution of their population even in very remote epochs. Under the pressure of the vast sediment layer, the tiny siliceous formations have sometimes caked into a hard, stony mass. In this case scientists must apply special techniques—such as taking thin sections and using corrosive materials—before they can separate and count the petrified skeletons.

The two American oceanographers, Glass and Heezen, planned to use the radiolarians to find out whether the dipolar reversals of the magnetic field had actually affected the course of evolution. But other scientists tried to dissuade them from making the attempt, since it seemed unlikely that the oceanographers would find what they were seeking. Most scientists felt that even if the temporary loss of the magnetic shield had profoundly affected life on the land, its effects would not have extended to the ocean floor. They believed that the earth's atmosphere must have functioned as a cushion to shield the earth from the solar wind, so that even if a few protons of solar plasma had reached the earth, the particles could not have penetrated the ocean waters. If the atmosphere could be considered dense enough to halt radiation, then the sea water must offer an even more impenetrable barrier. Thus other scientists warned Glass and Heezen that not one particle of solar wind could ever have penetrated to the sediments they intended to study. The ocean seemed to be the last place to look for evidence of the biological effects of a dipolar reversal.

Fortunately the two oceanographers flew in the teeth of the evidence and carried out their expedition anyhow. It seems a minor miracle that they were granted the funds they needed for their enterprise, when it was considered doomed from the beginning. It was also a stroke of good fortune that the two oceanographers chose to conduct their drilling in the Indian Ocean.

Glass and Heezen knew about the experiments with volcanic rocks that had been conducted by their geophysicist colleagues. Thus they knew that the last of the dipolar reversals of the magnetic field had occurred around 700,000 years ago. Since that time, the earth's magnetic north has been located at the point we are accustomed to calling the North Pole. Thus 700,000 years ago, the earth experienced the loss of its magnetic shield. For at least 1,000 years (and probably for several thousand) the solar wind beat down on our upper atmosphere, producing greatly increased quantities of radiant isotopes. The reversal of the magnetic poles directly preceding this one had occurred around one million years earlier. Continuing to count backward, reversals of the poles have been dated at 1,800,000 years earlier, at 2,000,000 years, 2,600,000, 2,900,000, 3,200,000, 3,500,000.

For several reasons, the two oceanographers chose the Indian Ocean as the site of their expedition. For one thing, the various sediment layers of this area were unusually thin. For whatever cause, the layers have settled and compressed more tightly here than elsewhere. The great advantage of taking drill samples here in the Indian Ocean lay in the fact that the drill could bring up several layers at a time. That is, it could gather material from several geological epochs during which a dipolar reversal of the magnetic field had taken place. We have noted that a period of one million years is normally compressed into a sediment layer from five to fifty-five yards thick. In the Indian Ocean, the drill was able to take a sample about 8½ yards in thickness which contained sediments from a period of 4,000,000 years. During this period, the magnetic poles had been reversed no fewer than eight times. Each time the two Americans took a sample, they could search in eight separate areas of the sample for the biological effects of a dipolar reversal!

Glass and Heezen were successful almost from the start. The two men scanned each layer for the meteoric compass needles which revealed when a dipolar reversal had occurred; then they examined the sediment above and below this point, searching for fossilized microorganisms. The organisms they found were primarily radiolarians; but the sediment also contained other forms of invertebrate marine life. When the oceanographers had finished classifying and counting the various animal populations, they were astounded by the results. The behavior of the earth's magnetic shield had apparently sealed the fate of no less than twelve separate species! Moreover, the magnetic shield had not only sentenced species to extinction; in several cases, a *new* species seemed to have sprung up at the time of a dipolar reversal. All twelve species had survived at least two of the reversals; most had survived five or more. But the time came when they vanished as abruptly as they had appeared; and their disappearance coincided with the disappearance of the magnetic field.

Scientists are notoriously skeptical beings. The discovery made by the two oceanographers did not really constitute final and irrefutable proof that the reversals of the geomagnetic poles had affected the course of evolution. On the other hand, even the most skeptical scientist was bound to sit up and take notice. The expedition had turned out to be more revealing than anyone—including Glass and Heezen themselves—could possibly have hoped.

But why were the two oceanographers so successful? What about that supposedly "irrefutable" argument that radiation could not possibly have penetrated to the depths of the sea?

The two American oceanographers have not attempted to explain the phenomenon. They confined themselves to a curt statement to the effect that whatever the true explanation, the study of fossilized marine organisms had for the first time enabled scientists to establish a connection between evolutionary crises and reversals of the earth's magnetic poles. Probably the explanation of this phenomenon relates to the fact that the solar wind itself did not trigger the evolutionary crises, which instead must have been caused by the radioactive isotopes created by the solar wind. We cannot say for certain how the C^{14} and other isotopes penetrated the ocean depths; probably they traveled downward by means of the food chain which, as we have already noted, extends to the depths of the sea.

Thus the sudden unpredictable occurrence of a reversal of the magnetic poles has always initiated a new chapter in the history of life on earth. If our cosmic observer were to watch the drama unfolding on the earth after the loss of its magnetosphere, he would probably see something very similar to the scenes described in this chapter. Life would never have evolved to this point in such a "brief" span of time if cosmic forces had not intervened in the course of evolution. That is, life would never have raced from the one-celled organism to man himself if radiation had not hastened the process. The periodic loss of the magnetic field not only *accelerated* the course of evolution; it also caused actual evolutionary "leaps"—completely new life forms with new modes of adaptation. We ourselves would never have existed if it were not for that giant invisible sphere which shelters us in space. We would never have become what we are today and will be in the future, if it were not for the strange pulse rate of that sphere, which beats only once in hundreds of thousands of years.

We should not forget that the magnetic "pulse beat" which occurred 700,000 years ago will undoubtedly not be the last. The farther back geophysicists have gone into the history of fossilized magnetism, the more dipolar reversals they have been able to date. During the past 76,000,000 years, at least 170 such reversals have occurred! We have no reason to believe that in our time the cycle has suddenly ended. On the contrary, indications are that reversals of the earth's magnetic poles will continue indefinitely. Of course, the next of these world-shaking events may not occur for a very long time yet. Perhaps by then the human race will have become extinct, and will therefore suffer no ill effects. But we have no way of knowing that this will be the case.

Russian scientists have completed a statistical analysis of the time intervals separating the various dipolar reversals. They have discovered that throughout the earth's history, these intervals have been steadily decreasing. We do not yet know the cause of this gradual decrease. Five hundred million years ago, the reversals of the magnetic poles occurred only once every 10,000,000 to 20,000,000 years. But 200,000,000 years ago, they normally occurred once every 1,000,000 years. During the past 20,000,000 years, the dipolar reversals have been separated by intervals of only 250,000 years. During this recent phase, two dipolar reversals have taken place within 10,000 years of each other. Commenting on these findings of their Russian colleagues, Glass and Heezen laconically remarked that the next dipolar reversal seems due in the very near future. No one can predict when the next reversal of the magnetic poles will actually occur. But it may sober us all a little to reflect how this event seems likely to affect the human race. After all, we long ago achieved our optimum adjustment to our environment. For thousands of years our species has been the undisputed master of the earth.

We should keep one other thing in mind regarding the evolutionary process. All life forms which have ever existed on earth, including man, are unique. None can ever be repeated. If history should begin again from the very first day the earth came into being, it would never reproduce the same effects; 3,000,000 or

4,000,000 years after life began, things would look very different from the way they do today. Even if the earth were the very same earth, and even if all the initial circumstances were exactly the same "the second time around"—nevertheless the course of evolution could never be exactly repeated. On this second earth, we would never be born again in the form we have today; nor would any other of the countless life forms on earth develop again. Too many random factors contributed to the development of each species born on this planet; too many unique combinations determined which of all of life's inherent possibilities would actually become reality. We have already noted that of all the possibilities inherent in life at its inception, only a tiny fraction have ever been realized on earth.

As an example of the uniqueness of all life, let us consider Neanderthal man, that competitor of our own ancestors, whom these ancestors wiped out some 30,000 or 40,000 years ago. Many scientists now believe that Neanderthal man possessed all the rudiments of civilization: the techniques of creating fire and stone tools, wall painting, and the burial of the dead. Probably our own ancestors learned and adopted all of these practices from Neanderthal man, along with certain basic religious concepts. What was he like, this man whose brain was substantially larger than that of his competitor, the father of modern man? Who was this man whose culture was vastly superior to that of his contemporaries? The structure of his lower jaw reveals that he did not possess the faculty of speech. Many scholars have studied the culture of our now extinct competitor and noted the remarkable size of his brain. They have seriously discussed the possibility that Neanderthal man may have developed his own form of communication—perhaps even thought transference! But we will never know the truth. Neanderthal man represents one of the countless possibilities of life which was never fully developed on this planet. His brain was quite different from ours. But we will never know how he experienced this world, or what he made of it all.

Cosmic Bull's-Eyes

WHEN Glass and Heezen took their drill samples from the sea, they found everything they had hoped to find —and something more. Besides the radiolarian skeletons and the cosmic meteor dust, their samples also contained tiny glasslike particles. Beneath the microscope these particles looked like broken chips of glass—only greenish-black rather than transparent. Many had the shape of teardrops, others looked like tiny buttons, still others resembled cut and polished spheres. But the most remarkable thing about these bits of vitreous dust was their location. They were found in every one of the yard-long drill samples taken from the Indian Ocean—but in only one layer of the samples. This layer marked the last reversal of the magnetic poles, which had taken place 700,000 years ago!

For a time no one could understand where the glasslike particles had come from or how they could be related to a dipolar reversal. But the unusual shape of the particles soon put investigators on the right track. Mineralogists had long been familiar with these remarkable-looking vitreous stones, although they had never before seen the stones in this "miniature edition." They were used to finding stones of this sort the size of pebbles or even of pigeon's eggs. The stones were known as "tektites." Apart from being thousands of times smaller, the microscopic droplets and beads found under the sea looked exactly like normal tektites. They also had the same chemical composition as the tektites, and mineralogists believed they must have been produced in much the same way. Thus the tiny chips were dubbed "microtektites."

Once the vitreous particles had been identified as tektites, scientists realized why they might have been concentrated in a single sediment layer. The tektites were rather a special case. For many years scientists had agreed that these strange stones were produced by collisions between the earth and the giant cosmic rock fragments known as meteors.

During the last few years scientists have finally uncovered conclusive evidence regarding the true origin of tektites. Before this evidence was found, mineralogists had been puzzled by the "riddle of the tektites" for almost two hundred years. The most famous tektites were the "moldavites" of Europe, named for the region where they were dicovered, the Moldau Valley of Bohemia. Because of their shiny vitreous appearance, these stones had been used for centuries in Moldavian jewelry. But it was precisely their vitreous character that caused mineralogists the greatest headaches, for the "glassiness" proved that the moldavites had hardened from a molten substance. If they had hardened from a molten mass, they had to be volcanic in origin.

Unfortunately, geologists unanimously agreed that volcanoes had never existed in that region of Bohemia. The baffled mineralogists finally resorted to desperate measures; they suggested that long ago a glassworks had existed in this area; the moldavites must be the fragments of bottles which had been broken during production. At the beginning of the nineteenth century, people even called the stones "bouteille," or "bottle," stones. But there was one fly in the ointment: No former glassworks had been mentioned in any of the historical documents of that region. Moreover, the glassworks must have been an exceptionally large place, since the greenish stones were scattered over an area comprising hundreds of square miles.

In the middle of the nineteenth century, reports of similar stones began pouring in from the far corners of the earth. A French geologist brought back some of the stones from Indochina. Shortly thereafter, Charles Darwin made his famous trip around the world, in the course of which he discovered more of the stones in Australia. Other finds were reported in the Dutch East Indies and in the Philippines. At first the stones were named after their supposed places of origin: indochinites, philippinites, australites, etc. But more and more of the stones were found all over southeast Asia. It became clear that all the stones, wherever they might be found, were chemically and mineralogically identical. Around the turn of the century scientists began to refer to all the stones, including the moldavites, simply as "tektites." (See Illustration 22.)

During the twentieth century, tektites were found in two additional regions, near the Ivory Coast of West Africa, and in Texas. Everywhere the problem was the same: The stones were clearly vitreous in composition and must have solidified from a molten mass. Yet nowhere in any of these regions could geologists find evidence of earlier volcanic activity.

The first step toward solving the mystery was made by German physicist Wolfgang Gentner, director of the Max Planck Institute of Nuclear Physics in Heidelberg. In 1959 Gentner hit on the idea of applying the isotope method to date tektites from all the various regions of the earth. He and his colleagues made an astounding discovery. No matter how many of the stones they succeeded in dating, they invariably ended up with one of four dates. Each of these dates was characteristic of one specific geographical region. All the tektites from southern Asia—regardless of whether they came from Indochina, the Philippines, Borneo, Java, or Australia—were 700,000 years old. West African tektites always tested out as around a million years old. All the European moldavites were 14,600,000 years old. The oldest stones were those from Texas, which were measured at 34,000,000 years. The age of the stones never varied a whit from these figures; every stone tested fell into one of the four groups. To this very day, every tektite found on earth has belonged to one of these groups.

Scientists realized that all the tektites belonging to a single group must have been formed simultaneously, by a single event. Moreover, this "event" had obviously occurred at least four times in the earth's history. What could possibly have happened? It must have been a major disaster; after all, the most recent of the mysterious events had scattered countless bits of molten glass all the way from the southern tip of Australia north to the Chinese mainland. Clearly only a mighty force could have formed such huge quantities of tektites and flung them over widespread areas. A number of researchers began to suspect that the rocks might be molten chips of giant meteors which had crashed into the earth on at least four separate occasions. A quick glance at the moon's craters shows us what huge rock fragments sometimes wander about in space, and what can happen when a heavenly body collides with one of these wandering rocks. The earth is considerably larger than the moon; therefore it offers a much larger target. Logically, our planet must have been struck by meteors far more often than the moon. (Although the earth's force of gravity is also far greater than that of the moon, gravity actually plays no role in attracting meteors or meteorites, since meteors do not really "fall" onto the earth. Rather the two bodies suffer a head-on collision because their two orbits happen to cross paths.) To be sure, the earth is not dotted with craters like the moon. When meteors enter the earth's atmosphere, they encounter air resistance and rapidly heat up; soon they explode from the heat and are shattered into many harmless fragments. But on rare occasions—every few hundred thousand or million years—the pattern is broken; the earth collides with a meteor that does not completely blow up in the atmosphere. Both the earth and the meteor are traveling at breakneck speed, so that the remains of the meteor strike the earth with devastating impact.

The craters formed by these collisions are slowly effaced by time. Unlike the moon, the earth has an atmosphere; it has wind and weather to wear down the crater's walls and fill in the depression at its center. All the same, it sometimes takes millions of years to remove all traces of a large crater. Thus scientists hoped they might still be able to find evidence that the tektite swarms were meteoric in origin; all they had to do was to search the areas where the tektites had been found, in the hope of discovering traces of old craters.

It was Gentner and his colleagues who solved the problem of the tektites. The Heidelberg physicists recalled the Nördlinger Ries, that shallow basin lying between the Frankish and Swabian ranges of the Jura mountains. Nowadays the German town of Nördlingen lies in this basin. The basin forms an almost perfect circle with a diameter of about 12½ miles (Illustrations 23 and 24). For a long time geologists had been puzzled by this extraordinary formation, which is totally unrelated to the geological structure of the surrounding area. During the 1920s, stone fragments were found near Ulm and Augsburg, sixty miles south of the Ries; other fragments turned up in other directions over a wide area. The mineralogical and chemical composition of these rocks showed that they must all have come from the Ries. Scientists wondered whether the Ries could once have formed the mouth of a giant volcanic crater. Vitreous masses of molten material had been found at the foot of the rocks surrounding the basin, and the rocks themselves formed a circular wall like the wall of a crater. At first this wall appeared to substantiate the theory that the basin might represent the mouth of a former volcano. But geologists soon realized that this could not be the true explanation; the rocks surrounding the depression were not volcanic.

The Nördlinger Ries lies about 186 miles west of the region where the moldavites were found. Despite the distance separating the two points, the little group of Heidelberg physicists led by Gentner decided to attempt a bold experiment. In 1962 they began to date the vitreous molten masses of the Ries. The startling result: All the glass fragments were precisely 14,600,000 years old. That is, the samples taken from the Ries were exactly the same age as the Bohemian tektites found 186 miles away! The chain of evidence was now complete.

Following this discovery, geologists began intensive investigations of the Ries area. They discovered widespread damage and traces of a sudden intense rise in temperature in the earth and rock beneath the basin. There is no longer any doubt that over 14,000,000 years ago a giant meteor struck the earth at this point. The huge chunk of cosmic rock must have had a diameter of six-tenths of a mile. The meteor was halted by the earth's crust, but the impact produced enough heat to create an explosion that released as much energy as several hundred hydrogen bombs put together! Showers of molten rock were hurled all the way to Bohemia 186 miles to the east, forming what we now know as moldavites.

The molten stones were not hurled indiscriminately in all directions; they all turned up in a relatively small region of Bohemia. Scientists were not surprised to discover that the molten showers had all fallen in a single area. The meteor struck the earth while traveling at a speed of from twelve to nineteen miles per second. According to the laws of aerodynamics, a stone mass this size which is traveling at such a speed when it strikes the earth at an oblique angle must create an area of vacuum in its wake. This vacuum would have acted like a giant vacuum cleaner, sucking up all the glowing liquid rock at the meteor's point of impact. In other words, most of the molten minerals must have shot straight into the air like a white-hot fountain; then the fountain curved around in a huge arc and fell back to earth 186 miles to the east.

Encouraged by their success, Gentner and his colleagues set out to track down the crater which theoretically ought to have been located near the West African tektites. They found it—in the shape of circular Lake Bosumtwi in Ghana. Although the lake was now filled with water, careful investigation revealed that its basin was actually a meteorite crater with a diameter of 4.3 miles. Here too the tektites were found about 186 miles from the actual point of impact.

On the other hand, scientists have not yet discovered the craters near the Texan and South Asian tektites. It is particularly difficult to explain why they have not found the Asian crater. The tektites of this region are scattered all the way from Australia to the Philippines and China, and scientists have calculated that only a meteor several miles wide could possibly have strewn tektites over so vast an area. The total mass of the tektites produced by the meteor crash has been estimated at 250,000,000 tons. Moreover, the disaster which took place here 700,000 years ago must have affected the entire earth. The crater produced by such a massive collision should not be too difficult to find. In addition, this meteor collision was the *last* of the four "events" to occur, so that the crater cannot yet have been worn away by wind and weather. Even if the crater were on the ocean floor, sophisticated modern oceanographic methods ought to have traced it long before this. Many scientists, including Gentner himself, suspect that this crater may be hidden beneath the miles-deep ice of Antarctica. Some time during the next few years, an expedition will set out for the South Pole to explore this possibility.

Meanwhile, we are left with the fact that the most recent of the reversals of the geomagnetic poles took place around the same time that our planet collided with a giant meteor. The two events may well be related. Scientists still know very little about the probable "dynamo" in the earth's liquid core and how it may act to maintain the magnetic field. Nevertheless, the collision with a meteor might easily have disrupted the even flow of the rotating metallic liquid at the earth's core, which presumably serves as the armature of this dynamo.

Oceanographers found microtektites in the sediment layers of the Indian Ocean. They found the stones in only one sediment layer: the layer formed during the last of the dipolar reversals of the magnetic field. Since this discovery, many scientists have regarded it as proven fact that the reversals of the magnetic poles are related to collisions with giant meteors. The meteors are like shots fired from space which strike their target head on. As they strike, these "cosmic bull's-eyes" cause the evolution of life to jump forward. The accidental nature of these collisions underlines the random character of the evolutionary process and the uniqueness of all life forms. The collisions indirectly determine which species will profit from the increased number of mutations, and which will simply perish.

We have noted the findings of Russian scholars that the frequency of dipolar reversals has been steadily increasing. At first glance this fact may seem to contradict our former statement that collisions with meteors occur entirely by accident. This steady increase in the number of dipolar reversals may appear to suggest that some sort of law controls the incidence of collisions between meteors and the earth. Such is not really the case, since we must keep in mind the fact that not all regions of space are the same. Our entire solar system is traveling through space at around nineteen miles per second; our earth constantly moves from one region of the Milky Way to another. During the past few million years of our flight through space, we may have entered a corner of the Milky Way in which meteors and meteorites are more plentiful than in those regions of the galaxy we formerly traversed.

Somewhere out in the depths of space, a giant rock seems to

be traveling aimlessly through the universe. It weighs hundreds of billions of tons. Today it may still be billions of miles from the earth. But it is already traveling on a course which one day—10,000 or 100,000 years from now—will cross the path of the earth. The collision between them will shake our entire planet; earthquakes and floods will devastate the land. Above all, the collision will disrupt the dynamo in the earth's core, collapsing the magnetic field. Once again the evolution of life will leap forward. The existing order of life will be destroyed; hitherto unknown life forms will transform the face of our planet. In the past these events have happened over and over. Why should the future be different?

The Universe Has Its Metabolism Too

EVEN after Gentner's discovery of the meteor craters, some scientists still believed that the tektites might actually have fallen from the moon. This hypothesis is not as far-fetched as it may sound. The tektites must have traveled through the air at great speed, since otherwise they could not have landed so far from the meteor's point of impact. But the tektites are rather small stones, not massive enough to have survived a flight through the earth's atmosphere traveling at such speeds. Aerodynamic calculations indicate that the air friction at this speed would have raised the temperature of the tektites to the melting point; they would simply have melted away in the air.

If the tektites had been traveling from the moon, the situation would have been different, for in this case they would not have been moving so fast. The moon's gravitational field is so weak that comparatively slow-moving objects can free themselves from the moon's influence and be drawn into the atmosphere of the earth. If the tektites had traveled from the moon, their surface would have been heated to the melting point, but only the tiniest fraction of their surface would actually have melted away. A few years ago the American aerodynamics researcher Dean Roden Chapman conducted some fascinating experiments with tektites. (It was Chapman who designed the legendary U-2 reconnaissance plane, one of which was shot down over Russia on May 1, 1960.) The American scientist heated natural and synthetic tektites in his wind tunnel, raising the temperature to the melting point. Then he subjected the tektites to various wind velocities in order to see what speed would produce the special blistered and striped surface markings characteristic of all genuine tektites. Having discovered this speed, he revealed the results to his colleagues. They were impressed by Chapman's findings; he had hit on exactly the speed at which an object would be traveling if it had been drawn from an orbit in space and entered the atmosphere of the earth!

At this point most nonscientists will probably be wondering what could have broken the stones loose from the moon's surface in the first place. After all, something had to propel them out into space before they could be attracted by the earth's gravitational field. Scientists had no difficulty in answering this question. Because of its weak gravitational pull, the moon exercises so little control over objects on its surface that a man on the moon with a long-barreled rifle could easily shoot a small satellite out into space. The satellite would not fall back to the moon; it would simply go into orbit *around* the moon. The same man with the same gun could even shoot an object so far that it would be immediately attracted into the earth's atmosphere. The moon has no atmosphere to protect it from meteors which crash into its surface. Thus the ensuing explosions produce great quantities of energy—more than enough energy to break off chunks of the moon and hurl them out into space. Scientists have estimated how much matter is lost by the moon during its collisions with meteors. To be sure, the moon also acquires matter by absorbing the meteors themselves; but scientists believe that the moon loses far more than it gains in this transaction. As the moon continues to orbit the earth, it releases matter at various points along the earth's orbit—matter which sooner or later is absorbed into the earth's atmosphere.

At one point Americans actually began to refer to the tektites as "moon stones." But despite Chapman's work, scientists are virtually certain that the tektites could not have come to us from the moon. Mineralogists believe the tektites to be molten material from the earth's own crust. Once again it was Gentner and his research team who succeeded in resolving the whole question. The Heidelberg team investigated the chemical composition of the "moon stones," subjecting them to every conceivable test, and then compared the composition of the tektites with the composition of rocks in the meteoric craters. The stones were identical in composition to material taken from the earth's crust; they were clearly not of lunar origin. Thus the "moon stone" theory was finally discredited.

The Heidelberg researchers were incredibly thorough in their investigations. They even performed a chemical analysis of the gas bubbles sealed inside the tektites—bubbles only a few thousandths of an inch in diameter. The scientists compared the gases contained in these bubbles with the composition of the earth's atmosphere and found their composition to be identical in every detail. Therefore scientists now believe that earth rocks are melted at the meteor's point of impact; these liquid stones are hurled several miles straight up into the earth's atmosphere. As they break out of the atmosphere, they encounter the chill of space and harden into round or droplike shapes before falling back to earth. As they reenter the atmosphere they heat up again, becoming the stones we know as tektites. The riddle of the tektites seems to have been solved for good.

The tektites definitely did not come from the moon. All the

same, they called attention to the fact that our planet must be filled with extraterrestrial matter. Somehow scientists had never seriously considered this possibility before. To be sure, they knew that each year an estimated 5,000,000 metric tons of meteor dust rain down upon the earth. But if we convert this figure into the quantity of dust that falls at each individual point on the earth's surface, we find that only one-millionth of a gram falls on each square centimeter of surface. This seemed such a minimal quantity that scientists thought little about the question of extraterrestrial matter. But when research scientists tried to unravel the riddle of the tektites, they began to calculate the various effects of collisions between tektites and the moon, and suddenly realized that meteor dust was not the only extraterrestrial matter that was falling onto the earth from out of the sky.

Tektites are composed of terrestrial matter hurled several miles into space. But this fact does not completely invalidate all aspects of the "moon stone" theory. Meteors *do* frequently crash into the moon, and the resultant explosions really *do* hurl large quantities of lunar matter into space. The earth's gravity field actually does attract the lunar matter, which sooner or later ends up on the earth's surface. This process has been going on for billions of years, so that by now the surface of our planet must be almost buried in lunar matter. There can be no doubt that at some time in our lives, each of us has held a moon rock in his hand. Unfortunately, we cannot tell the difference between an earth rock and a moon rock just by looking.

Moon rocks and meteor dust cover the earth's surface, but they are not our only extraterrestrial visitors. Day after day chunks of metal and rock enter our atmosphere in the shape of "shooting stars." Each of these falling meteors or meteorites consists of "unearthly" matter from other parts of our solar system. A constant stream of matter has been flowing toward this planet ever since it first came into being. But not all this matter comes to us from somewhere inside our own solar system.

We have noted that comets "belong" to particular solar systems in the sense that they orbit a particular sun. But as we mentioned before, comets describe highly eccentric orbits in the form of an extended ellipse. It may take a comet thousands of years to complete a single circuit around its sun, and at the most remote points of their orbits comets may be two or three light-years away from their suns. At these remote points, the comets are already traveling in the border regions of neighboring solar systems, where they sometimes cross paths with comets of these neighboring systems. At such times, two comets may actually exchange orbits. That is, each comet begins to orbit a new sun. Thus a constant exchange of matter takes place at the border regions of neighboring solar systems. When a comet approaches the center of its system, it falls under the influence of the various massive bodies in this region, including its own sun.

A planet sometimes diverts a comet from its course, drawing the comet into its own atmosphere, where the comet explodes into hundreds of pieces. In this way the matter composing a comet may eventually end up on the earth. This comet may belong to our own solar system; or it may belong to one of our "neighbors."

How fascinating to realize that the surface of our planet must be strewn with matter from other solar systems! In recent years scientists have discovered that the exchange of matter takes place on a truly cosmic scale. Illustration 25 demonstrates this point. During their flight through the universe, the galaxies shown in this photograph came very close together and exercised a mutual gravitational attraction. Each galaxy began to attract great masses of interstellar dust and meteoric material from all the other galaxies in the group. This cosmic debris streamed out from the galaxies and began to form a sort of "bridge" between them. The bridge now extends over hundreds of thousands of light-years. What we see in this photograph is a sort of cosmic "metabolism."

It seems appropriate to apply a biological term to this astronomical phenomenon. The word "metabolism" literally applies to this case. The matter composing the universe is continually being "thrown beyond" (from *meta*, "beyond," and *ballein*, "to throw") its momentary narrow sphere. As it travels through the universe, this matter tears down and builds up all objects, providing the raw material for new creation. This "metabolism" resembles the metabolic processes of organic life.

The astrophotographers who snapped the picture in Illustration 25 were not recording a unique event; our Milky Way too contains matter from other milky ways. During the past few billion years, a certain quantity of extragalactic matter has continually filtered down to the surface of the earth, where it has been slowly accumulating on the earth's crust. Many scientists estimate that the entire crust may now be largely composed of extraterrestrial matter.

The fact that the surface of our earth is covered with cosmic matter may help us to understand two things about ourselves. First, we on earth are completely unable to distinguish extraterrestrial or extragalactic matter from the matter native to our planet. This fact substantiates a view long cherished by our astrophysicists, that the entire universe is composed of the same basic substance. Second, we are now in a position to see how relative the term "extraterrestrial" really is. The ground on which we are standing is "unearthly." But what does this fact imply? In the past we have associated the word "extraterrestrial" with something exceptional and even alien. This attitude reflects a certain basic prejudice, the same prejudice which we noted at the beginning of this book: the view that the earth represents an isolated unit. Clearly, the matter composing the universe does not share our prejudice against what is seemingly "foreign."

Some of us may feel that the discovery of the cosmic exchange of matter has taken a little of the magic out of the word "extraterrestrial." But we have gained something which should make up for this loss—the insight that at least in a material sense, the earth and the universe are not altogether alien to each other. The earth does not constitute a "foreign body" in space. Instead, the earth is the child of the universe; the matter composing our planet came out of the depths of space.

There is another reason why the cosmic exchange of matter may be compared to a metabolic process. Cosmic metabolism was indirectly responsible for our own metabolism; it established one basic precondition for the development of life. Next we will learn the story of the relationship between the universal exchange of matter and our own biological processes. It will be the last story in this book. Like our other tales, it offers an example of the indissoluble unity of all phenomena—the unity between microcosm and macrocosm.

We Are Such Stuff as Stars Are Made Of

OUR last story begins in the year 1944. This tale has an all-star cast and tells how our Milky Way Galaxy with its hundreds of billions of suns might easily have remained a dead island in space. We ourselves might never have been born—if our galaxy had not had a spiral shape!

For years astronomers have known that there were several different types of galaxies. Some galaxies are rather diffuse-looking lens-shaped formations—the so-called "elliptical" galaxies. These elliptical galaxies contain approximately the same number of stars as all spiral galaxies, including our own Milky Way Galaxy, and both types of galaxies occupy about the same volume. In fact, elliptical galaxies are virtually identical to spiral galaxies in all but one respect: The elliptical formations have no spiral arms (see Illustrations 27 and 28). Scientists believe that no life forms like our own can exist in the elliptical galaxies, which appear to have remained in a stage of development antecedent to the development of life as we know it. In order to understand this situation properly, we must begin at the beginning. In 1944 the German-American astronomer Walter Baade published a treatise on "Star Populations." For many years Baade had been working at the famed Mt. Wilson observatory, investigating the spectral lines of stars outside our own galaxy. (The surface temperature and chemical composition of stars can be determined from the spectral lines present.) While engaged in these investigations, Baade observed that all spiral galaxies seemed to contain two distinct types of stars. He referred to these types as "star populations"—a phrase meaning something like "star species."

Baade called the first star species "Population I." This group consisted of relatively young, very hot stars, which because of their heat emit a bluish-white light. They are "only" a few million years old—100,000,000 years at most. For the most part these stars are concentrated in the spiral arms of all the far-off milky ways.

The second type of star composes Star Population II, which exhibits just the opposite properties. These stars are much cooler than those of the first group; thus their light tends to be reddish in color. They are several billion years old. Baade found that these stars were fairly evenly distributed throughout the darker areas *between* the bright spiral arms.

One other property distinguished the two star populations, chemical composition. Investigation of the spectral lines revealed that the young, hot stars were composed of from one percent to four percent heavy elements, including metals. On the other hand, the Population II stars seemed to consist chiefly of densely compressed hydrogen, containing only a tenth or even a hundredth as much of the heavier elements.

Since 1944, astronomers have been busy investigating Baade's two star populations, and now believe that they understand what causes the difference between the two types of stars. Baade's initial discovery also set astronomers to working on various allied problems. Their work has opened our eyes to a basic fact about the cosmos—a fact which no one had even dreamed of before. For some time everyone had known that the stars we see in the sky vary in age and in their distance from the earth. But eventually astronomers learned that stars actually belong to various successive star generations!

Astronomers did not arrive at this startling conclusion until some time after Baade's discovery. First they asked themselves a different question: Why do all the young stars congregate in the *arms* of the spiral galaxy? Their studies revealed that this occurs for the same reason that most newborn children are found in maternity hospitals: They are born there! Most new stars develop in the spiral arms of their galaxy because stars form through the condensation of interstellar matter, and the spiral arms contain the largest quantities of this interstellar matter.

This fact helped to explain a phenomenon which had long puzzled astronomers, the stability of spiral-shaped galaxies. These gigantic formations rotate on their own axes, but they do not turn as a wagon wheel does, i.e., as a single rigid unit. Instead, the various components of the system rotate at different speeds, largely depending on their distance from the center of the galaxy. The rotation of the galaxy is extremely complicated. At the center of the galaxy, the rotation begins at zero and increases very rapidly, almost as if the galaxy *were* a solid wheel. At a certain distance from the center, the rotation slows down again, then speeds up, finally peaks, and declines from there on. That is, the rotational velocity of our galaxy follows a complex curve.

There is a good deal of dispute over the precise velocity of various suns in the galaxy. The estimates of astronomers are derived not from direct observation but from some working model of galactic rotation which appears to be a good approximation of reality. Our sun's velocity is generally estimated as less than 160 miles per second. Suns at the edge of the galaxy rotate more slowly, perhaps at around 135 miles per second. Despite their speed, these "border" suns may take some 400,000,000 years or more to complete a single orbit around their common galactic center. Our sun has a far shorter distance to travel, and therefore takes only about half this time to complete one orbit—about 250,000,000 years.

Up to a point the behavior of galaxies seemed entirely consonant with the laws of mechanics. But when astronomers began to compute the varying speeds of the different bodies within a spiral galaxy, they realized that after only two complete rotations around their common center, the spiral arms ought to have simply "been wound up." That is, the galaxy could not possibly have maintained its shape over the course of time. Yet astronomers knew that the oldest existing spiral galaxies must have completed at least twenty rotations since the universe began, and these galaxies nevertheless still retained the characteristic spiral shape which had given them their name. The astronomers were baffled as to how to explain this apparent contradiction of mechanical laws.

Although for some time the problem seemed insoluble, some astronomers believe that they have now found the true explanation. Their theory is based on the recent discovery of intragalactic magnetic fields contained in all spiral galaxies, including our own Milky Way. Some unknown force causes magnetic fields to develop at the center of a galaxy; the fields then spread out to the borders of the entire stellar system. The lines of magnetic force radiate outward rather like the spokes of a giant wheel. But the galaxy which contains them is constantly rotating at high speed; thus the force fields do not radiate outward in a straight line. As the galaxy turns, the lines of force are turned aside and bent into an arc. Astronomers do not know all the details of this process. But apparently the curving lines of magnetic force act more or less as the supportive "spine" of the spiral arms.

Magnetic fields might offer a satisfactory explanation for the stability of the giant rotating galactic "wheels." For billions of years all parts of the galaxy have been turning right on cue. A formation governed by simple mechanical laws alone would long since have lost its shape. But the spiral lines of magnetic force fields can never be bent out of shape. These fields are invisible and nonmaterial; only their effects are visible. A large percentage of the hydrogen gas in any stellar system is ionized (i.e., the particles of the gas are electrically charged), so that the hydrogen atoms are subject to magnetic influence. Thus the hydrogen may tend to arrange itself along the spiral path of the magnetic fields.

There is considerable dispute among astronomers regarding the formation of the spiral arms of spiral galaxies. Many astronomers believe that magnetic fields do not actually create the spiral arms, but may merely help to *maintain* the spiral structure once it has been established. At some point invisible clouds of neutral hydrogen begin to contract under self-gravitation; later the gravitational fields of these clouds cause them to contract faster and faster. These "primary clouds" form the nuclei of newborn stars. The developing suns pass through the same stages as our own sun (see the chapter called "Portrait of a Star"). Differential rotation may stretch out the clouds into long spiral-shaped streamers. As the density of the spiral clouds increases, stars begin to form. The newborn stars throw off immense quantities of ultraviolet radiation, which ionizes all the neutral hydrogen in their vicinity. The now ionized hydrogen may bring the magnetic field lines along with it as it continues to contract, so that the magnetic field helps to maintain the spiral structure of the stellar system.

Thus the stars of a spiral galaxy are not actually arranged in a spiral pattern. (We have noted that a merely "mechanical" formation of this sort could never maintain its pattern over the course of time.) The spiral formation is merely apparent. Astronomers now have reason to believe that the stars of a spiral galaxy are more or less equally distributed throughout the galaxy. The spiral appearance results from the fact that the youngest and brightest stars are concentrated in certain areas—the areas where the gas from which they form is most abundant.

Indications are that the dark-looking areas between the spiral arms are also filled with stars. Possibly there are as many stars in the dark areas as in the bright ones. These areas appear dark to us because the stars here are older and shine less brightly. Moreover, these regions contain so-called "neutron stars." The recently discovered neutron stars exhibit a number of remarkable properties: For one thing, they are so faint that some of them may be virtually invisible! Soon we will discuss these strange bodies in greater detail.

Astronomers have discovered many things about the internal structure of a milky way. They now know why young stars congregate along the spiral arms. They have also learned that the young stars do not remain in the spiral arms throughout their lives. Unlike the thinly compressed ionized hydrogen gas, the stars are very massive; they can no longer be controlled by the galactic magnetic fields. As the giant galactic wheel slowly rotates, completing a single rotation every 250,000,000 years, the massive young stars slowly move out of the spiral arm regions. Meanwhile, newborn stars replace them in the bright spirals.

Baade's discovery in 1944 raised a number of questions, none of which has really been "answered"; all our present answers are merely hypothetical. One question astronomers now face relates to the chemical composition of stars. We noted that Baade's two star populations differ in chemical makeup. The old stars consist almost exclusively of hydrogen, the element from which they originally formed, whereas the young stars contain traces of many other, heavier elements. Where then did these elements come from?

In the chapter dealing with our sun, we briefly touched on this problem, mentioning that the nuclear fusion process in the sun's core produces helium as the end product. The helium represents a sort of "ash," the remains of the hydrogen consumed in the atomic process. Actually this "ash" constitutes an element in its own right—the second in the periodic table of the elements. In our chapter on the sun, we concerned ourselves only with the sun's energy production. But now we must consider in more detail the process by which the elements are produced.

Our world is composed of ninety-two natural elements. These elements might never have come into being—if it were not for the stars! All stars have their personal biography; they live through a definite life cycle. Throughout its life, every star must maintain a balance between two forces, the atomic energy in its core, which presses outward, and the gravitational force which presses in, holding the star together. Eventually a star uses up all the hydrogen fuel in its core; that is, the star changes all its hydrogen into helium. For a time the star continues to maintain its stability. Once the hydrogen in the core has been changed to helium, the atomic fire slowly spreads outward from the core, consuming the hydrogen in other regions of the star. But eventually the atomic fire eats its way near to the surface. Soon the outer layers of the star can no longer exert enough pressure to set off atomic reactions. The atomic furnace burns out.

Once the atomic furnace is extinguished, atomic energy ceases to press outward from the core. In the past this outward-expanding force had counteracted the star's gravitational force, preventing the gaseous mass from contracting further. But now the central regions of the star begin to contract again, once more building up great pressure and heat in the star's core. This central contraction releases enough energy so that the outer regions of the star become greatly extended, cool, and red-colored. The star is now in the so-called "red giant" stage. During this stage, the pressure in the central regions increases to more than 2 billion metric tons per square inch; the temperature rises to

15,000,000°C. In the past, this temperature and pressure set off atomic reactions at the core which consumed the star's hydrogen as fuel. Now there is no more hydrogen, only helium. Helium must be subjected to much higher pressures and temperatures than hydrogen before it will undergo fusion.

At this point no expanding force exists in the star to offset its contraction. It continues to contract until its temperature rises to more than 100,000,000°C, the critical temperature at which concentrated helium begins to "fuse." Atomic reactions begin again, preventing the further contraction of the star and temporarily stabilizing it. During this phase, some of the helium is gradually turned into carbon. Some of the helium briefly changes to beryllium, which rapidly decays, producing oxygen.

Eventually the helium nucleus of the star is consumed. Then the whole process repeats itself. The star contracts until it creates even higher temperatures than before, at which point the carbon atoms begin to be transformed into heavier elements. The carbon becomes neon and sodium. A complex chain reaction produces new helium nuclei which serve as "building blocks" for other elements. The chain reaction eventually creates magnesium, aluminum, sulfur, and calcium.

By this time the temperature of the core has risen to some 500,000,000°C. At this point the atomic processes taking place in the core become so complex that we cannot adequately describe them here. The atomic reactions produce so much energy that the various elements are constantly being torn down and built up again. Eventually the star exhausts all its store of nuclear energy. The star has long since been driven from the spiral arms where it was born and now moves through the darker regions of the galaxy. It has become what astronomers call a "white dwarf." Its extraordinary heat makes it look white. Yet despite this white heat, the star's light appears rather dim. During its successive stages of contraction, it has lost most of its volume and shrunk to the modest scale of a good-sized planet. It may now be no larger than Jupiter!

The "dwarf star" may have a diameter of little more than 60,000 miles. Confined within this sphere is the mass of an entire sun—a mass which once occupied at least a thousand times this volume! The core of a white dwarf star consists of extraordinarily dense or "degenerate" matter. A single cubic centimeter of this dense matter actually weighs many tons!

What happens next depends on the star's total mass. The critical mass is about 1.4 times the mass of our own sun. When a star like our sun, which weighs less than this critical mass, reaches the white dwarf stage, it continues to contract; but it triggers no new atomic reactions.

Until recently astronomers considered this the "normal" behavior of a star in its final phase of development. But in one sense this behavior seems anything but normal—in the sense that it is unproductive. A star of this kind passes through alternate periods of contraction and atomic fusion reactions, in the course of which it develops various heavier elements from the hydrogen of which it was originally composed. But this star never progresses beyond the production of nickel. That is, it produces only about one-quarter of the total number of elements. Moreover, certain elements created by such a star immediately break down again; among these are lithium and beryllium.

Thus these "normal" stars manufacture a limited number of elements. More importantly, even those elements it does produce remain trapped inside the star. The elements were created through a long and complex process. But what purpose can they serve if they stay buried forever in the cooling core of a white dwarf? It seems anything but "normal" that nature should expend so much effort in vain; these stars seem to represent a kind of dead end. It is as if the universe were being "deprived" of the elements produced in this sun's core.

But we set out to learn the source of the matter composing our earth and the other planets. Where do the ninety-two elements come from, those elements which combine to form our world and even our bodies? This question was answered by the

Indian astronomer Subrahmanyan Chandrasekhar, who discovered what would happen if the mass of a white dwarf exceeded 1.4 times the mass of our own sun. Chandrasekhar's computations revealed an astonishing fact. If a star's mass exceeds this critical mass, then at a certain point in its contraction after the red giant stage, its gravitational field becomes very intense. That is, the star's gravitation pulls more strongly on the matter composing the star until it actually destroys the atomic structure of the matter. In other words, long after it has reached the white dwarf stage, the star simply goes on contracting.

The "degenerate" matter of a white dwarf star is extraordinarily dense. The electron shells of its atoms have all broken down. But when a large star continues to contract, something else happens in addition to the breakdown of electron shells. There comes a point when the atomic nuclei themselves break down. The moment when this breakdown occurs is known as the "gravitational collapse." Before its collapse, the star is still as large as a good-sized planet. Then suddenly, in a split second the entire star crumbles away. In a moment it has a diameter of only six to twelve miles! This "implosion" generates temperatures of more than 3,000,000,000°C. It is impossible even to imagine the 15,000,000° at the center of our sun; during the gravitational collapse of a white dwarf, the temperature rises to two hundred times that of the sun's core. As the temperature reaches this point, an atomic explosion tears away one-tenth of the star's total mass, hurling this matter into space at speeds of up to 6,000 miles per second.

When a fixed star explodes in this manner, it produces a completely new star, a "supernova." For a few weeks a supernova may radiate the light of 200,000,000 suns. In an earlier chapter we discussed the occurrence of supernovas in our own and other milky ways (Illustration 26). In recent years astronomers discovered the cause of these huge stellar catastrophes: the gravitational collapse of a white dwarf which exceeds the critical mass.

Scientists did not learn all these facts about stars by studying them directly. At first they employed "model stars" in conjunction with computers. The models supplied astronomers with data about the various stages in the life history of a star. But scientists also have direct evidence that the "implosion" process —the gravitational collapse of a star under the weight of its own degenerate matter—takes place in outer space as well as inside a computer. This direct evidence has been provided by rocket telescopes specially designed to record X rays.

The explosion of a supernova leaves behind it a new star. The mass of this star still more or less equals that of our sun, but this entire mass is compressed into a sphere no more than six to twelve miles across. The matter composing the star now consists solely of densely packed neutrons. For this reason astronomers refer to it as a "neutron star." A single cubic centimeter of this neutron matter weighs several million tons! If we removed a piece of matter from the core of a neutron star—a piece no larger than a matchbox—and placed it on the earth's surface, it would immediately break through the earth's crust and fall all the way to the earth's core, where it would finally come to rest.

A star composed of such abnormally dense matter is bound to display a few other "eccentricities." Probably the strangest quality of a neutron star is its near-invisibility. Neutron stars are burning at temperatures of several billion degrees, but this brilliant light is still very faint. Anyone reading this sentence may imagine that the neutron stars are simply very small—no larger than a medium-sized asteroid—and therefore virtually invisible over long distances. But actually it is thought that if the original star had a high enough mass, then the resulting neutron star would in fact be literally invisible. The mass of several suns has been packed into a tiny area. It is possible that the gravitational field of this mass may sometimes grow so powerful that not even photons of light (which enables us to see things) can escape it!

The English astronomer Fred Hoyle theorized that a neutron star may sometimes more or less close itself out of our universe. Neutron stars that have succeeded in closing themselves out of the universe are called "black holes," but their existence is merely theoretical; no one has actually succeeded in observing one. These black holes would betray their presence only by their powerful gravitational attraction, which could be felt enormous distances away. Perhaps many of these invisible stars are strewn throughout the universe. One day man may travel among the stars. If he does so, the black holes may be lying in wait for his spaceships—rather as rocks or reefs hidden under the sea once lay in wait to wreck passing ships. What an uncanny death trap that would be: an invisible star whose gravitation would suck in any spaceship that inadvertently approached too near!

From the earth, astronomers cannot detect the gravitational field of a neutron star. But computers have revealed that neutron stars also emit powerful X rays. X rays are completely absorbed upon entering the earth's atmosphere, so that earthbound laboratories cannot accurately record them. But the American scientist Herbert Friedman succeeded in equipping a rocket with an X-ray telescope; he shot the rocket above the earth's atmosphere, where it remained for several minutes. In this way Friedman was able to locate a number of tiny X-ray sources deep in space. One of these sources lay in the famous Crab Nebula, which, as we already mentioned, actually represents the exploding cloud of a supernova. Friedman's findings indicate that neutron stars are more than mere fantasies dreamed up by our electronic computers. But although we have no reason to doubt other facts computers have told us concerning neutron stars, astronomers are not yet able to understand everything the computers have revealed.

Computers report that the neutron star does not represent the final stage in a star's biography. The star may have a diameter of little more than six miles. At this point, various processes which we do not yet understand begin to generate new, very heavy elementary particles. After this brief interruption, the process of contraction sets in once more. But this time it simply continues. The neutron star shrinks away to a purely *mathematical* point; that is, it becomes a sheer abstraction.

Let us not attempt to understand what finally becomes of this star. Suffice it to say that no force exists capable of halting the star's withdrawal into itself, so that eventually it passes into a realm which is no longer accessible to mathematical calculation. In the course of gravitational collapse, the star ejected a large percentage of the elements it had created in its core. After its implosion, the star simply exits from the stage. In some incomprehensible manner, it literally disappears from the universe.

We have now answered our two original questions: How does a star composed of the primal matter of creation, hydrogen gas, generate all the other elements? And, how do these elements escape from the star so that they may be used in new formations? The explosion of a supernova destroys a star which has fulfilled its purpose. At the same time, the explosion marks the first stage in the birth of a new star. This new star belongs to a different generation which in no way represents a mere carbon copy of the previous one. The second-generation stars are not born from pure hydrogen. Instead, they form from clouds of interstellar dust. This dust contains heavier elements produced by earlier stars, elements released into space by the explosion of a supernova. The cloud of the Crab Nebula (Illustration 8) represents more than the remnants of a dying star; it also supplies the matter for new birth.

Directly or indirectly, all these findings resulted from the work reported by Baade in 1944. Since that time astronomers have learned that far more than two "populations" exist among the stars. We now know that many successive star generations have come and gone, each of them forming from the matter provided by the previous generations. All the matter existing in the various spiral galaxies has been used over and over; it alternately condenses into stars and is released again into space.

Gradually the stars have manufactured all the heavy elements in the periodic table of the elements, including the heaviest of all, uranium.

Intragalactic magnetic fields bring about the local concentration of hydrogen in the spiral arms of the galaxy. To see the importance of this fact, we need only examine the elliptical galaxies which have no such spiral arms. We have not yet discovered the cause of this structural difference between spiral and elliptical galaxies. But we do know that elliptical galaxies do not possess the hydrogen arms which appear to act as the "germ cells" of new stars. This fact seems to explain why elliptical galaxies do not evolve, that is, they do not give birth to new stars. Elliptical galaxies contain only superannuated stars consisting almost entirely of hydrogen and helium. Spectroscopic analysis has revealed no trace of heavier elements.

Thus elliptical galaxies lack most of the elements which function as the building blocks of nature as we know it. Unlike our galaxy and all other spiral galaxies, elliptical galaxies apparently lack the capacity for further development. They cannot provide raw material for the formation of planets; nor would they be able to furnish a planet with soil, water, and air. Thus elliptical galaxies will never progress to the final logical step, the manufacture of organic molecules. The evolutionary process begins on the chemical and continues on the biological plane; but elliptical galaxies will never give birth to life as we know it.

Like all spiral galaxies, an elliptical galaxy contains hundreds of billions of suns. It occupies as much space as our own Milky Way Galaxy, and its history is as ancient as that of other galaxies. It differs from our own galaxy in only one respect: It cannot produce life resembling that on earth. So far it has not even succeeded in manufacturing the ninety-two elements which compose our world.

All the matter around us—the substances that make up the earth, the plants which grow on it, the animals which inhabit it, the very matter composing our own bodies—every atom of this matter was conceived in the glowing heart of some unknown sun. The sun existed unspeakably long ago; it belonged to a star generation which is now long gone. If certain vast cosmic events had not taken place, nothing in our everyday world would now exist. The universe used an entire milky way with its hundreds of billions of suns in order to create the commonplace objects that surround us.

Children of the Universe

In the beginning was hydrogen. Besides hydrogen, nothing else existed but the laws of nature and space itself—unimaginable quantities of space. The world was born from an immense cloud of hydrogen gas which began to contract under the influence of its own gravitation. The process of contraction began some 10 to 15 billion years ago. Our entire Milky Way Galaxy, including the earth and ourselves, developed from this primal cloud. Meanwhile the same thing was happening at billions of other points in the universe, where various condensing clouds were giving birth to the billions of galaxies we observe in the heavens today.

Throughout countless millions of years, the giant cloud continued to contract toward its own center of gravity. It was impossible for all the hydrogen atoms to aim at precisely the same point. That is, the cloud was still too diffuse to have a single clearly defined midpoint. The contracting atoms tended to graze or collide with each other. Meanwhile the entire formation increased the speed of the "carousel" motion we have already observed in other heavenly bodies. For a time the contracting cloud had possessed a roughly spherical shape. The quickening of the carousel motion intensified the centrifugal force which was tending to modify the shape of the cloud. Over the course of hundreds of millions of years, centrifugal force slowly flattened out the entire formation: The cloud took on the shape of a giant discus with a diameter of more than 100,000 light-years.

In the earliest phase of its history—before it had really begun to rotate and flatten out—something else began to happen to the giant cloud. Hydrogen gas began to form local concentrations in certain areas of the cloud. These concentrations of hydrogen gas were the seeds of the first stars. We can still identify the most ancient suns of our stellar system. They developed when our Milky Way Galaxy had not yet developed the shape it has today; that is, they were born when our galaxy still had a roughly spherical shape. Thus these ancient stars form characteristic patterns in space. They are not distributed in the same way as more "modern" stars. The oldest stars form the so-called "globular clusters." These remarkable clusters contain many hundreds of thousands, or even many millions of stars arranged in a globular pattern. This pattern has an average diameter of "only" about 200 light-years (Illustration 29). Unlike all other stars in our galaxy, these stars do not lie in the flat, disk-shaped plane of the Milky Way. Thus far astronomers have discovered 119 such globular clusters in our galaxy, none of which lies in the plane of the galaxy as a whole. Instead, they are evenly distributed on all sides of the center of the galaxy, composing a sphere. Their spherical distribution around the center of the galaxy reveals the archaic character of these formations.

Astronomers believe that these ancient stars mark the space

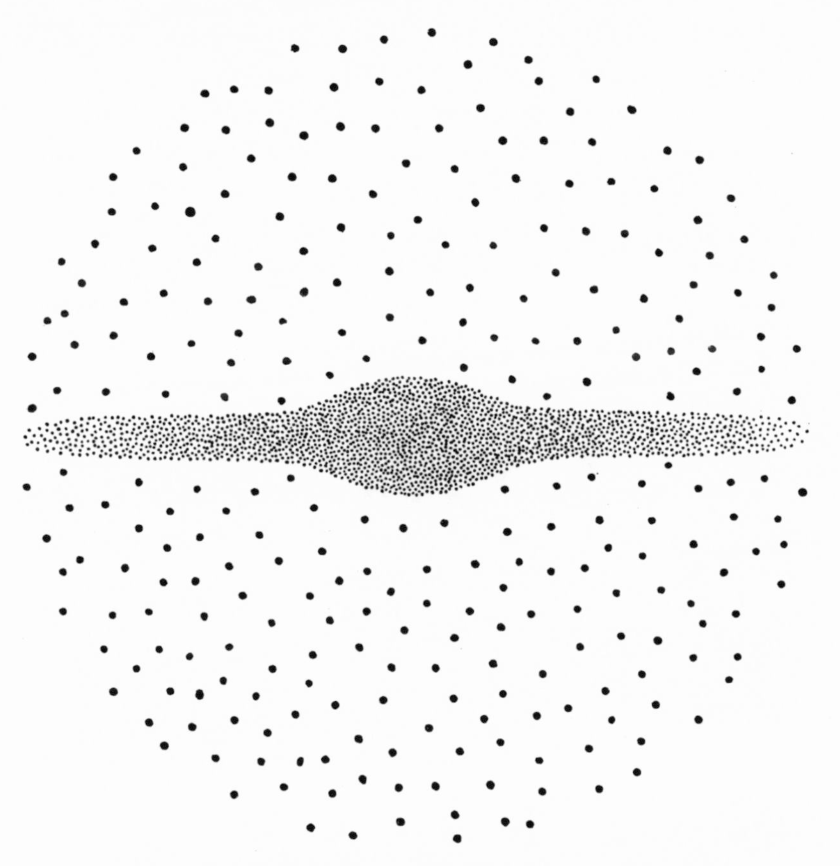

Average distribution of the globular clusters in our galaxy.

occupied by our Milky Way Galaxy when it was still a spherical cloud of gas. The archaic members of our galaxy exhibit several peculiarities which support the belief that they developed before the rest of the present Milky Way Galaxy. For one thing, the suns belonging to a globular cluster are in fact the oldest stars in our stellar system. Their age has been estimated at somewhere between 6 and 10 billion years. They appear to consist of pure hydrogen and helium, with no admixture of heavier elements. Moreover, globular clusters move along their own independent orbits; that is, they do not rotate with the rest of the galaxy. There is only one possible explanation for this independent motion: These ancient stars came into being *before* our galaxy began to rotate!

Astronomers have observed the globular clusters of other nearby galaxies and found that these clusters exhibit the same properties as those in our own galaxy. Our nearest neighbor, the Andromeda Galaxy, is only about 2,000,000 light-years away, and therefore can be examined in considerable detail. Astronomers have already detected some 200 globular clusters in this galaxy, all of which seem in every way identical to those in our own Milky Way system.

Gravity and centrifugal force combined to form our galaxy, giving it the flat disk shape typical of all spiral galaxies. Apparently the production of stars began in earnest only after our galaxy had acquired this characteristic shape. To be sure, the globular clusters contain one or two billion stars. But apart from these, all the stars in our galaxy are concentrated in the disk-shaped space occupied by the modern Milky Way.

Once our galaxy had acquired its discus shape, the first generation of stars came into being. These suns were composed of pure hydrogen, the lightest of all elements and the primal matter of the creation. In time the hydrogen condensed sufficiently to generate atomic reactions in the core of each star. Once atomic reactions began to occur, the stars were well on their way to producing the rest of the ninety-two elements composing our present world.

Billions of years passed. Innumerable suns transformed their hydrogen into helium, then changed the helium into carbon, oxygen, and other new elements. Once more, billions of years went by. All the stars were destroyed in enormous explosions. In exploding, they released new elements into space in the form of fine interstellar dust, which gave birth to a new generation of stars. This new generation was enriched by the heavier elements; in turn, it generated still heavier ones. And so the process continued, with the destruction of each star generation creating the raw material for higher development.

Around 5 billion years ago, cosmic evolution entered its second stage. New suns were being born which for the first time contained traces of every one of the ninety-two natural elements. The suns began to form planets from these ninety-two elements. Of course, the only planets we have actually observed are those in our solar system. Even the nearest fixed stars are too far away for us to see their planets. But after all, our Milky Way Galaxy contains some 100 billion stars very like our sun, and it would be strange indeed if only one among all these stars had developed planets. Even if we are very conservative and assume that only one out of every 100,000 suns developed its own planetary system, this would mean that 1,000,000 solar systems like our own existed in this galaxy alone!

The same argument holds true for the development of life. How can we assume that, of all the planets in our Milky Way Galaxy, only one should have given birth to life and consciousness? Yet many people still make this assumption, reflecting the medieval illusion of geocentricity. They may have discarded the belief that our planet is the physical or astronomical hub of the universe, but they still believe that in the entire cosmos, only the earth could have generated consciousness or intelligence. For them, the earth is still the center of the universe. Actually, this attitude represents a throwback to the old Ptolemaic system. Logic indicates that all spiral galaxies must be teeming with conscious life. At this very moment throughout countless billions of galaxies, innumerable varieties of creatures are brooding over the mysteries of the universe.

The human imagination cannot begin to picture what life forms may exist elsewhere in space. Probably we would not understand these creatures even if we should ever meet them face to face. We have adapted to a specific environment which has narrowed our range of possible growth. Although we may be sure that we are not alone in the cosmos, we cannot really imagine who our companions may be.

About 4 billion years ago, when the earth was not yet quite one billion years old, solar radiation began to cause simple chemical compounds on the earth's surface to merge into larger molecular units. The molecular units became the basic building blocks of organic life. Once these building blocks were created, the earth could never again be the same.

The next step in terrestrial evolution occurred about one billion years later, when the first living cells came into being. They were highly organized structures capable of reproducing themselves through cell division. They were free to move about as they chose, and although dependent on their environment, they formed separate organisms, each with its own metabolic functions. The first individuals had been born.

Another million years went by. It took the microscopic organisms all this time to populate the primeval oceans. As they spread, they developed a multiplicity of new life forms adapted to various environments. Some of these organisms even proved capable of adapting to changing conditions. And meanwhile, the very presence of these countless microorganisms began to modify the surface of the earth.

Until this time, there had been virtually no oxygen in the earth's atmosphere. But then the tiny microorganisms in the primeval seas began to produce oxygen. These organisms were small, but they existed in great numbers all over the earth and were able to permanently alter the chemical composition of the atmosphere. All the oxygen we breathe today was originally created by these creatures. Oxygen provided living organisms with a new source of energy; they could now move about more actively and develop in new directions.

The oxygen atmosphere accelerated the tempo of evolution. Only 500,000,000 years later, living creatures left the seas and began to conquer the land. After another 200,000,000 years, the dinosaurs ruled the world. One hundred million years sped by, and nature "invented" the warm-blooded organism. Warm blood enabled an organism to move about freely, regardless of the temperature outside its body. Only 50,000,000 years later, many bird and mammal species had spread throughout the earth.

It might well have seemed that life had now evolved to its highest possible level. Then once again a new force entered the scene. We do not know just when the phenomenon of consciousness first came into being; in any case, this phenomenon began to appear among the highest living organisms. Before this time, animals had been capable of flight, of hunger, and of caring for their young. From now on, they felt fear, curiosity, and affection. One million years later, certain organisms had developed such a high level of consciousness that they were actually aware of their own existence! Man had come into being–and with man, all that we know as "civilization."

No one witnessed the history of evolution. Man has only recently discovered the existence of the evolutionary process, the basic principles of which were described little more than a century ago. We are only now coming to recognize that we are the product of an immeasurably long course of development. Also, we are just learning to see that we are not the *final* product of this development. At this very moment life is evolving toward a future we cannot even begin to imagine.

In the long history of evolution, consciousness has existed for only the briefest of moments. One instant ago life crossed the threshold from dull sensation into genuine self-awareness. Only one living creature on earth has succeeded in crossing this threshold–man himself. For thousands of years human beings vainly attempted to understand the mysteries taking place far above them in the heavens. With the advent of recorded history, they became aware of another mystery, themselves. Man and his immediate surroundings are as enigmatic as anything in the depths of space.

At first man approached the world subjectively. He believed his planet to be the stable hub of the universe with sun, moon, and stars revolving around it. Later he adopted a critical stance toward himself and his environment, seeking to view the world "objectively," i.e., independently of himself. The scientific method was born. The first product of the scientific method was the "Copernican crisis." Human beings began to feel that they had been marooned on a tiny particle of dust drifting forlornly through a hostile universe. This nightmarish image helped to mold the human psyche for 400 years. But once again there is a change in the air. In the past, science disillusioned us. Now the same science is helping us to wake up from our long nightmare.

We have not, after all, been abandoned in a universe whose inhuman beauty has nothing to do with us. We are not drifting through a vacuum unrelated to our earth. Nor are we tolerated only because of our sheer insignificance. Immeasurable time and immeasurable space gave birth to the life on our planet, and they preserve it still.

Countless billions of suns were born and died to create the matter that now composes our world. We ourselves are formed from matter created in the stars. The vastness of space established the preconditions for life on this and countless other planets. Earth and sun, moon and solar system all helped to bring us into being. Cosmic forces maintain the stability of our frail familiar world. The cosmos itself is the ground and origin of our existence.

We have taken a brief look at the new picture that scientists are fashioning of our world. Suddenly our solar system appears to be traveling through a different universe. The cosmos is no longer cold and hostile; it is *our* universe. It brought us forth and it maintains our being. We are its creatures. This fact should give us faith as we travel on our way—even if no one can tell us where the journey ends.

Index

Hoimar von Ditfurth

Hoimar von Ditfurth was born in Berlin in 1921. He has been a scientific journalist for many years, and a professor of psychiatry and neurology at the Universities of Würzburg and Heidelberg. In Germany he has won acclaim for his television series "Science in Cross Section," and in 1967 won the Golden Camera award for distinguished work in TV broadcasting. He has also been awarded prizes by a number of German scientific societies.